D0471523

MAN'S CONQUEST OF SPACE

BY WILLIAM R. SHELTON

Foreword by James E. Webb
Former Administrator, National Aeronautics and Space Administration, and Member of the Society's Board of Trustees

PREPARED BY NATIONAL GEOGRAPHIC SPECIAL PUBLICATIONS DIVISION
Robert L. Breeden, Chief

NATIONAL GEOGRAPHIC SOCIETY
WASHINGTON, D.C.

Melvin M. Payne, President
Melville Bell Grosvenor, Editor-in-Chief
Gilbert M. Grosvenor, Editor

Man's Conquest of Space

By William R. Shelton

Published by
The National Geographic Society
Melvin M. Payne, *President*
Melville Bell Grosvenor, *Editor-in-Chief*
Gilbert M. Grosvenor, *Editor*
Kenneth F. Weaver, *Consulting Editor*

Prepared by
The Special Publications Division
Robert L. Breeden, *Editor*
Donald J. Crump, *Associate Editor*
Philip B. Silcott, *Manuscript Editor*
Leon M. Larson, *Assistant to the Editor*
Johanna G. Farren, *Research and Style*
Geraldine Linder, Linda Seeman, Tee Loftin Snell, Betty Strauss, *Research*
Richard M. Crum, Ronald M. Fisher, William R. Gray, Jr., Gerald S. Snyder, *Picture Legends*
Robert Cumming, *Consultant to the author on Space Sciences*
Luba Balko, Margaret S. Dean, Carol Oakes, Judy Strong, Sandra Turner, Barbara Walker, *Editorial Assistants*

Illustrations and Design
Donald J. Crump, *Picture Editor*
Geraldine Linder, Michael E. Long, *Assistant Picture Editors*
Joseph A. Taney, *Art Director*
Josephine B. Bolt, *Assistant Art Director*
Robert W. Messer, *Production*

Staff for the Third Edition
Philip B. Silcott, *Managing Editor;* Louis de la Haba, *Manuscript Editor;* Toby Turner Shelton, *Assistant to the Author;* Donald J. Crump, *Picture Editor;* Toni Eugene, Penelope A. Loeffler, *Research;* John S. Graham, *Picture Legends;* Ursula Perrin, *Art Director;* George V. White, *Production;* Raja D. Murshed, Nancy W. Glaser, *Production Assistants;* John R. Metcalfe, *Engraving and Printing;* Mary G. Burns, Jane H. Buxton, Marta Isabel Coons, Suzanne J. Jacobson, Sandra L. Matthews, Marilyn L. Wilbur, *Staff Assistants;* Martha K. Hightower, Virginia S. Thompson, *Index*

 Library of Congress Catalog Card Number 79-173307; Standard Book Number 87044-069-1
Third Edition 1974

MOON SHIP Eagle *lifts toward rendezvous with the Apollo 11 command module* Columbia *after its historic lunar landing, July 20, 1969. The craft carries the first two men to set foot on the moon: Neil A. Armstrong and Edwin E. Aldrin, Jr. Overleaf: Echo 1 satellite streaks across the dense, starry center of the Milky Way. Page 1: Footprint in the Sea of Tranquillity may remain a million years or more. Front endsheet: Craters Aristarchus (upper left) and Herodotus (upper right), photographed in 1971 during the Apollo 15 mission, mark the moon's near side. Back endsheet: Craters never seen from earth scar the far side, photographed from Apollo 11.*

MICHAEL COLLINS, NASA (RIGHT); OVERLEAF: HAROLD ABLES, U. S. NAVAL OBSERVATORY, FLAGSTAFF, ARIZONA; PAGE 1: NEIL A. ARMSTRONG, NASA; FRONT ENDSHEET: ALFRED M. WORDEN, NASA; BACK ENDSHEET: MICHAEL COLLINS, NASA.

FOREWORD

Before boarding *Friendship 7* for lift-off on February 20, 1962, Astronaut John H. Glenn, Jr., carefully tucked a small scrap of blue, brown, and green cloth in with his gear. This piece of material—a miniature National Geographic Society flag—traveled almost five hours in space with the first American to orbit earth. Six months after his flight, the personable Glenn presented the flag to Dr. Melville Bell Grosvenor, the Society's President and Editor, as a salute to the National Geographic's "pioneering contributions to space research" and its "many years of strong support to those men who seek to explore the unknown."

Indeed, your Society has played an integral role in the exploration of space. It has supported many scientists in their endeavors and published their accounts in the NATIONAL GEOGRAPHIC. The Society's interest in space began early with the experiments of such pioneers as Alexander Graham Bell and Robert Goddard; it developed further with such achievements as the ascent of the *Explorer II* balloon 13.71 miles into the stratosphere in 1935—a flight that held the altitude record for 21 years. Since the advent of the Space Age in 1957, the NATIONAL GEOGRAPHIC has continued to keep its members abreast of the manned and unmanned space ventures that have paved the path leading to the moon. In keeping with this tradition, the Society has prepared this third edition of *Man's Conquest of Space*, by William R. Shelton, who has written about the space effort since its beginnings. His first edition, published in 1968, won the coveted first-place book award of the Aviation/Space Writers Association.

To meet the challenge of reaching the moon—a stepping-stone in the conquest of space—Congress in 1958 established the National Aeronautics and Space Administration. Dr. Hugh L. Dryden, a Trustee of the Society and a man supremely confident of our future in space, became Deputy Administrator. In the 1960's he said: "Before the end of this decade man will launch his greatest voyage of discovery, a journey whose magnitude and implications for the human race dwarf any high adventure of the past." NASA coordinated America's immense resources and talents in developing a sound space program. Beginning with the early exciting days of the first satellites, the Nation's space effort achieved spectacular successes—the manned Mercury and Gemini flights, the faithful weather and communications satellites, the Mars and Jupiter probes, the Apollo lunar voyages, and the Skylab missions.

Now that man has reached the moon—what next? Space presents a limitless future and a limitless challenge; the only ingredients needed to meet them are man's imagination, his desire, and his inventiveness. A quarter of a century ago, few would have been bold enough to predict seriously that man would walk on the moon before 1970. And now, with the impetus already generated and the growing momentum of the space program, who dares prophesy the monumental strides we will take in the coming decades. The world of space holds vast promise for the service of mankind, and it is a world we have only begun to explore.

JAMES E. WEBB
Former Administrator, National
Aeronautics and Space Administration

CONTENTS

Apollo 11 Astronaut Edwin Aldrin deploys aluminum foil to trap high-velocity atomic particles in the solar wind. Such moon-based experiments yield clues to the solar system's origin.

NEIL A. ARMSTRONG, NASA

1/ MAN INVADES SPACE, THE LIMITLESS FRONTIER

Over the centuries the resolute spirit of the explorer has met every challenge it has encountered—scorching deserts and expanses of ice, plunging canyons and defiant peaks, the ocean depths and the lofty reaches of the atmosphere. But now man confronts what John F. Kennedy called "this new ocean" of space, the only limitless physical frontier he has ever faced. Never before has he been able to cut loose the shackles of earth to ascend toward a realm essentially unknown and without boundary.

In August of 1955, before Sputnik 1 and the dawn of this new age, the NATIONAL GEOGRAPHIC called the beckoning ocean of space "the last, the greatest, and the most dangerous frontier of all." How man fares in it may determine not only the nature of his future life on earth but, in a larger sense, whether he has a significant destiny away from his home planet.

How fast man should penetrate this new realm raises new and sometimes baffling questions. Man could relate to most previous frontiers in a personal way; often he could become a participant and explorer himself. This is not yet true of space. So far, only highly trained and select specialists have emerged beyond the atmosphere or landed on the plainlike *marias* and majestic highlands of the moon. We have also dispatched exquisitely fashioned robots to explore neighboring planets and investigate, from a promising new perspective, earth itself. During the brief history of the Space Age, man the individual has largely had to be content with his role as spectator.

As every schoolchild has read—from Novosibirsk in Siberia to Great Falls in Montana—the Russians astonished the world by launching the first artificial satellite on October 4, 1957. On that fateful morning on the grassy steppes of south-central Russia, the countdown inexorably reached *tri . . . dva . . . odin.* Then a gleaming white rocket—in a crescendo of power and sound—climbed into the sky, spun out its cottony contrail, and, as the distant sound funneled out, arched into an

FIRST TO ACHIEVE MAN'S DREAM *of space flight, Russian Cosmonaut Yuri A. Gagarin orbited earth on April 12, 1961, just 23 days before U. S. Astronaut Alan B. Shepard, Jr., rocketed into space. Major Gagarin, called the Columbus of the Cosmos by his countrymen, died in a plane crash in 1968.*

IGOR SNEGIREV, NOVOSTI

JOHN BRYSON, LIFE © TIME INC. (BELOW); SOVFOTO

elliptical orbit of the earth that would take it 560 miles into space at its highest point.

About 100 minutes after launch, the orbital path of the 184-pound Sputnik brought it again over Russia; excitement mounted at the rocket base. The leader of the Soviet team was a brilliant engineer, Sergei Pavlovich Korolev, a pioneer whom the Russians thought of as "a man who could put rivets in his dreams."

Now he carefully checked Soviet tracking instruments. In his headset, he could hear a steady "beep beep"—a series of crisp, clear, high-pitched notes that told him Sputnik 1 was securely in orbit.

Later, half a world away, millions more heard the unchallengeable beep of the strange satellite on American radio and television.

Commentators struggled to explain the phenomenon: The satellite had arched into space at just such a precise speed and altitude on just such a heading parallel to the surface of the globe that its centrifugal force was in balance with gravity trying to pull it back to earth.

The *New York Times,* struck with the significance of the flight of Sputnik 1, said in an editorial: "The creature who descended from a tree or crawled out of a cave a few thousand years ago is now on the eve of incredible journeys."

As the Soviet Union erupted in a display of

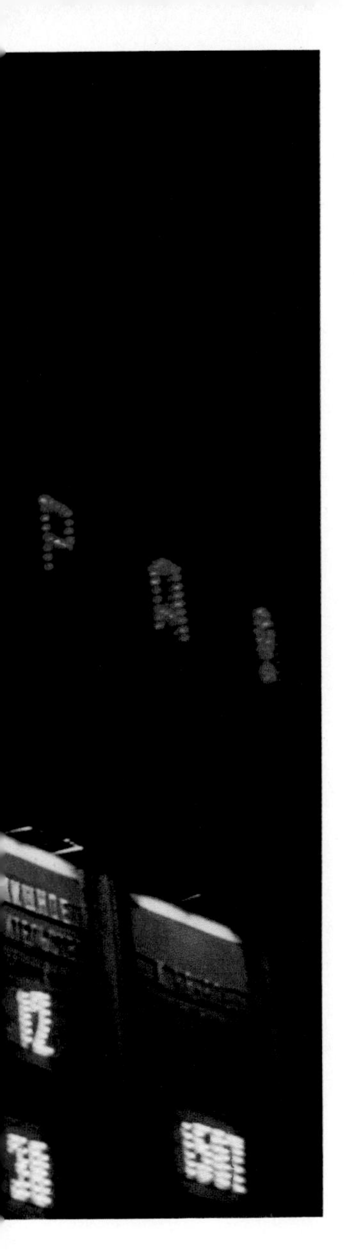

LIGHTS *on the Central Telegraph Office in Moscow celebrate the dawning of the Space Age in 1957, and the 40th anniversary of the Bolshevik Revolution. Above portraits of Russian leaders, a rocket soars toward a slogan that reads "U.S.S.R., Stronghold of Peace." Moscow amateur radio operator Valentin Vasilishchenko (below) picked up Sputnik's "beep beep" as it circled the earth. His family listened as he relayed information to ham operators worldwide.*

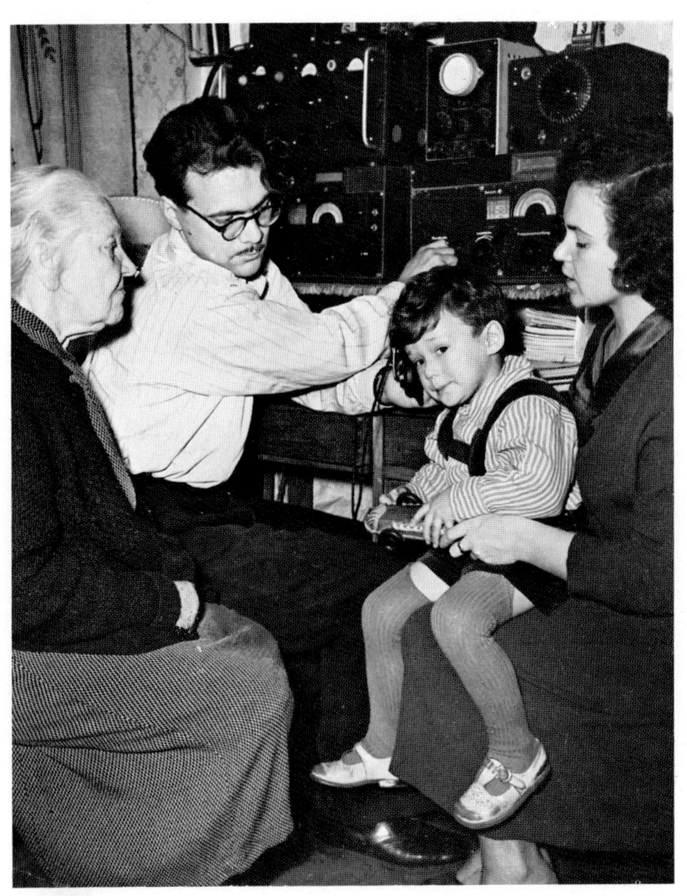

exhilaration, the prevailing mood in the United States was one of sober disappointment that our own Vanguard satellite, then awaiting launch at Cape Canaveral, Florida, had not led the way into space. President Dwight D. Eisenhower had announced that the United States, as part of the International Geophysical Year of 1957-58, would attempt to send a satellite into orbit, but the Russians had been first.

We had scarcely adjusted to the implications of Sputnik 1 when a second shoe fell. Just 30 days later, on November 3, 1957, the Soviet Union launched not simply an instrumented sphere, but a 1,120-pound payload. Sputnik 2 contained, besides a store of intricate scientific equipment, a small black-and-white dog named Laika.

American scientists realized at once that the Russian mastery of the art of rocketry was indeed formidable. And they knew that the achievement of supporting life in the alternating bake-freeze of space was a highly delicate and complex undertaking. Aside from the sophisticated life-support system the Soviets used, other instrumentation on board was designed to investigate the ionosphere, cosmic rays and other radiation, and the nature of the earth's magnetic field.

About a week after launch, signals from Laika's sensors stopped, indicating she had died. But the half-ton Sputnik 2 satellite and its rocket still remained in orbit three weeks later when those of us interested in seeing the long-awaited Vanguard launch assembled at Cape Canaveral.

The scene leading up to lift-off—T-minus-0—was as charged with drama as those days, hours, and minutes that toll off the approach of battle. The launch was a highly public affair; "research in a fishbowl," one reporter called it. Each day headlines shouted across the land that the countdown was on, then off, then on and off again.

I recall one night when I strolled down the beach across the Port Canaveral inlet from Pad 18A. The red and white lights of the Vanguard gantry danced in the dark waters. In the reflection, the rocket was transformed from a solidly poised stalagmite aimed at the heavens to a wavering stalactite pointing to the depths of the sea—a visual paradox that somehow matched my feelings of mingled hope and concern.

This was a December night, and even in Florida the beach felt cold and damp from the surf. Newsmen, sipping hot coffee, huddled near a line of driftwood fires blazing along the fringe of rustling palmettos and sea grapes. Missile buffs, or "bird watchers," mingled with them, occasionally joining in song.

Flickering lances of firelight outlined the bulks of trucks and moving vans that photographers had positioned on the beach as bases for their cameras. Shoreward, sea oats swayed in the wind, and in the dunes the light played on an incongruous battery of the tripods and telephoto lenses of the press—pointing toward Vanguard like the remorseless snouts of cannon.

The final countdown began at 1 a.m. on December 6. We heard little singing that night as we waited out a lead-colored dawn. Then through binoculars we saw frost building up on Vanguard's first stage as the sub-zero LOX (liquid oxygen) was pumped aboard. At 11:44 a.m. the rocket's umbilical, or service, cords dropped away. Launch was imminent. I strained for a clearer view of a moment of history.

Suddenly a splayed tongue of flame darted from the base of the rocket. Was this the fire of ignition—or something else? Vanguard began to rise. Then it stopped. It began to sink back. As it fell into its own launch ring, its thin skin ruptured. Instantly, a mass of incandescent red flame and boiling black smoke engulfed the rocket. The exploding fireball blossomed 70 feet high. In a few seconds it was all over—a national aspiration gone up in flame and smoke.

U. S. NAVY FROM NATIONAL ARCHIVES

FIERY EXPLOSION *consumes America's Vanguard rocket—and abruptly ends the country's first attempt to orbit a satellite. Vanguard's nose cone topples seconds after the blast on December 6, 1957, at Cape Canaveral. Above, technicians preparing for the launch effort position the nose cone above the rocket's 3.25-pound payload. Less than two months later, on January 31, Explorer 1 rode a flame-tipped rocket into an elliptical orbit, reaching 1,573 miles at its highest point, and blazed the trail for America's entry into the age of space.*

The intended satellite, which Soviet Premier Nikita S. Khrushchev was later to call an "orange," broadcast its cry not from the lofty throne of space but from the palmetto flats where grew thistle, swamp myrtle, and thyme.

As the news spread to a waiting Nation and world, it became obvious that the United States—on the surface at least—had suffered a catastrophe of enormous magnitude.

"Overnight," Dr. Wernher von Braun said later,

S-1312-A
S-1312-A

JUPITER-C ROCKET *towers above Army technicians (opposite) as the countdown begins for Explorer 1 on January 31, 1958. Evaporating liquid oxygen swirls around the feet of the men (below) as they fuel the rocket at Cape Canaveral in the stark glare of searchlights. At a blackboard, rocket expert Dr. Wernher von Braun, then Director, Development Operations Division of the Army Ballistic Missile Agency at Huntsville, Alabama, traces the satellite's trajectory after launching.*

"it became popular to question the bulwarks of our society, our public education system, our industrial strength, international policy, defense strategy and forces, the capability of our science and technology. Even the moral fiber of our people came under searching examination."

Many scientists warned that so great was the difference between the 1,120-pound Sputnik 2 with its life-supporting atmosphere and the tiny 3.25-pound Vanguard satellite, that it would be years before the United States could catch up. Pentagon officers pointed to a similar disparity in the relative sizes of booster rockets, which doubled as intercontinental ballistic missiles.

Against the background of pessimism in the days following Sputnik 2, the United States had pressed into service a unique team of rocket specialists to augment the Vanguard program with a second, separate project. Dr. Von Braun and his group of about a hundred former German rocketmen, then civilians working for the U. S. Army at Huntsville, Alabama, were instructed to prepare for the launching of an artificial satellite. Thus began one of the most brilliant chapters in the American conquest of space. Ever since the German rocketmen had been transported to the United States to live, Von Braun and his specialists

MAXWELL COPLAN, DESIGN PHOTOGRAPHERS INTERNATIONAL (ABOVE) AND U. S. ARMY

UE

N. R. FARBMAN, LIFE © TIME INC. (ABOVE) AND ROBERT PHILLIPS

SCANNING THE SKY *above New Mexico, a Smithsonian Institution Baker-Nunn tracking camera pinpoints Explorer 1. Jubilant scientists—Dr. William H. Pickering, Director of the Jet Propulsion Laboratory at Pasadena, California, Dr. James A. Van Allen, and Dr. Von Braun—hoist a mock-up of the final stage and payload. In the 112 days its batteries operated, the satellite transmitted cosmic ray, micrometeorite, and temperature data, and revealed a radiation zone later named for Van Allen.*

had proposed orbiting a satellite. Now, with the world watching, they had their chance.

On November 8, 1957, when Von Braun and Maj. Gen. John B. Medaris got the green light, Von Braun promised to rocket a satellite into space within 90 days.

On the night of January 31, many of us who had watched the exploding Vanguard's crimson glow stain the blockhouse walls assembled again at Cape Canaveral—this time to see if Explorer 1 could succeed where Vanguard had not. Officials allowed us inside the Cape area to a press site about a mile from the launch pad. When I first saw the Jupiter-C booster, it already radiated incredible luster inside its ring of searchlights. Through binoculars, its ice-encrusted form glistened like a pillar of cavern quartz caught in glaring light. The crisscrossing beams went on to pierce the blackness of the night sky, as if pointing

out the pathway to the stars. The entire Cape, with its ghostly array of antennas, theodolites, and towers, seemed to be in an eerie state of suspension. Photographers nervously checked their cameras every few minutes.

Just before 10 p.m. a foggy-throated warning horn sounded mournfully across the palmetto flats. At 10:38 the 18.13-pound satellite and the rocket's upper stages started slowly to whirl, then rapidly wound up until the top of the rocket was spinning at nearly 500 revolutions per minute, providing stability for the payload during flight.

Success seemed as irrevocable as the final tolling of the countdown. Suddenly, the bluish-white rocket spurted fire. A pink cloud of smoke and steam blossomed instantly. The rocket rose, gained speed, roaring out its power in great surges of sound that drowned out the muffled fragments of cheers. The rising flame-tipped stiletto pierced a cloud; then its light—redder now—reappeared above the clouds, coursing higher. Explorer 1 was on its way.

At that moment a friend of mine, CBS newscaster Charles Von Fremd, approached, his face and clothes wet with perspiration.

"That's the greatest thrill I've ever had in reporting," Chuck breathlessly exclaimed.

As Explorer 1 went into orbit, well within the

N.G.S. PHOTOGRAPHER DEAN CONGER (ABOVE); NASA

STRIDING PURPOSEFULLY *across Cape Canaveral's Pad 5, from transfer van to Redstone rocket, Mercury Astronaut Alan Shepard keeps a rendezvous with history on May 5, 1961. He carries an air conditioner to cool his pressure suit, and wears plastic overshoes to avoid tracking dirt into the space capsule. The suborbital flight, first U. S. manned mission, carried Shepard 302 miles to splashdown in the Atlantic. In an earlier simulated shot, Shepard (left) tests his equipment.*

three-month estimate of Von Braun, the rest of the country fully shared our jubilation; the United States had staked out its own claim in space. Explorer 1 was followed by the launch of Vanguard 1 on March 17, 1958. Now we, like the Russians, had dispatched two representatives to emerge beyond our lower atmosphere.

Explorer 1, during its 1,573-mile reach into space at the height of its elliptical orbit, revealed an earth-girdling band of trapped solar radiation. The band, considered then a potential hazard to astronauts, became known as the Van Allen belt, after Dr. James A. Van Allen, designer of the satellite's instrumentation. A later probe indicated a second belt, actually a similar, distinct zone within a single vast region of radiation. Vanguard 1, through deviations in its orbital path, confirmed earlier beliefs that the earth is slightly, almost imperceptibly, pear-shaped.

Our emotional, scientific, military, and economic commitment to space had become a fact. Congress and public called for a more ambitious program. On President Eisenhower's recommendation, Congress created a single federal space agency, the National Aeronautics and Space Administration. It began operation in October 1958.

NASA eventually acquired, in addition to thousands of scientists and technicians, many military and quasi-military organizations, facilities, and equipment. The Army was directed to transfer to the agency not only Von Braun and his 4,200-member group at Huntsville but also more than 100 million dollars' worth of facilities required to design, develop, test, build, and launch space vehicles. Not least of the rich legacy inherited by NASA were the detailed plans for the Saturn I, the first rocket especially designed by the U. S. for space flight.

The over-all space mission was broadly based and incorporated many aspects of exploration, from meteorology to communications, but its later directive to send a man into space was most eagerly embraced by the public. The first phase of the manned-flight effort was Project Mercury, a program to construct a vehicle to carry one astronaut. First, we would aim the craft like an artillery shell on a ballistic trajectory to take a man briefly into space. Then we would try something infinitely more difficult: We would attempt to send an astronaut into earth orbit.

NASA selected and began to train its first class of astronauts, undertook a massive construction program, and initiated a worldwide network of tracking stations for both unmanned and manned satellites. As the months passed, Project Mercury moved steadily toward its appointment with whatever destiny awaited the first man to venture into space. There were rumors that Russia, too, was preparing to send men into orbit. The question often arose: Who would blaze the trail, the United States or the U.S.S.R.?

Wherever I traveled during the first few months of 1961—at Langley Field, Virginia, home of the NASA Space Task Group, at the spacecraft factory of McDonnell Aircraft Corporation in St. Louis, and at Cape Canaveral itself—I heard nothing but

DOUGLAS MARTIN (OPPOSITE) AND NATIONAL GEOGRAPHIC PHOTOGRAPHER DEAN CONGER, BOTH FROM NASA

CHEERS *of spectators on Cocoa Beach (left) rise with Shepard as the Redstone rocket lifts off at the Cape. The 83-foot, 33-ton "bird," photographed through a 600-millimeter lens, appears closer than its actual distance of six miles. After the mission, Shepard, joined by the other six Mercury astronauts, received NASA's Distinguished Service Medal from President John F. Kennedy at the White House. The President inadvertently dropped the medal, quipped, "This decoration has gone from the ground up," and joined in the spacemen's laughter.*

confidence that America would lead the way. In February the country learned the names of the men chosen for special training for the debut in space—Navy Lt. Comdr. Alan B. Shepard, Jr., Marine Lt. Col. John H. Glenn, Jr., and Air Force Capt. Virgil I. Grissom. Triumph seemed assured.

But the honor of orbiting the first man was not to be ours. I was having breakfast in Chicago on April 12 when I learned that Russia had sent Maj. Yuri Alekseyevich Gagarin around the earth. To reach the Cape as quickly as possible, I took a jet to Tampa, then went by rented car across the peninsula to the Atlantic.

An hour after I reached my motel, where press cables were already being strung out for the impending Shepard flight, I had dinner with two of our astronauts, Donald K. Slayton and L. Gordon Cooper, Jr., and two of their physicians, Dr. William K. Douglas and Dr. Carmault B. Jackson, Jr. As we talked, we all agreed on one thing: We were deeply disappointed that the first man had not been an American. But as we got up to leave, "Deke" Slayton added a note of optimism.

"There's one good thing about Gagarin's flight," he said. "It proves there is no serious obstacle to a man going into space. Maybe we ought to be happy to know that it can be done."

Twenty-three days later, the press again entered the launch complex of Cape Canaveral, this time shortly before dawn, to witness what we hoped would be one of the great milestones of the decade. Clouds obscured half the sky, and as the sun rose above a gray bank of scud over the Atlantic, it tinted undulating wraiths of ground fog an incredible rosy gold. By 9:30 a.m. the countdown for *Freedom 7* had just four minutes to go.

"Well," said my friend Chuck Von Fremd nervously, "Alan is on his own now."

"Four . . . three . . . two . . . one . . . zero . . . Ignition . . . Lift-off!"

Shepard's calm, unhurried voice came to us. "This is 7 . . . Fuel is go . . . and the oxygen is go . . . Cabin pressure is holding at five point five."

The rising rocket plunged into a cloud, its flame appearing to perish like a match smothered in moisture. But the rocket reappeared, thrusting upward, gaining altitude. I vaguely heard cheers and applause.

"What a beautiful view!" Alan exclaimed as he reached maximum altitude of 116.5 miles. America's pioneer astronaut and his Mercury capsule performed flawlessly, and he later shared credit for the mission with the thousands of scientists, engineers, and technicians directly involved in the 15-minute, 302-mile suborbital space flight.

Slightly more than two months later, millions watched on television as "Gus" Grissom repeated Shepard's success—except for the accidental sinking of his capsule, *Liberty Bell 7,* after completion of the mission. Then in February of 1962, millions more were engrossed with the drama of John Glenn's three orbits of the earth in *Friendship 7.* When the smiling Marine colonel made his enthusiasm, sincerity, and sense of humor evident in Washington, New York, and other cities, Americans everywhere had a new light in their eyes and a new lift in their hearts.

"In the saddle of success," wrote Saul Pett of the Associated Press, "he rode loose and easy, and everyone found something to like."

So absorbing was this initial phase as man himself penetrated the cosmos that we scarcely realized how many records had fallen almost overnight. Before rockets carried man into space, a number of painstakingly prepared ventures had attempted to send him to ever greater altitudes and speeds. In 1935 the U. S. Army Air Corps and the National Geographic Society sent two balloonists, Capt. Albert W. Stevens and Capt. Orvil A. Anderson, to a record altitude of 13.71 miles. On May 4, 1961, the day before Alan Shepard's flight, the Navy sent Comdr. Malcolm D. Ross and Lt. Comdr. Victor A. Prather, Jr., in the Strato-Lab to a balloon record of 21.5 miles. The X-15 rocket plane and other experimental aircraft had often taken man to great heights, but the record altitude before manned space flight was the 25.85 miles reached by X-15 pilot Robert M. White on August 12, 1960.

Suddenly—befitting a major and compelling technological breakthrough—all these records were left far behind. The orbital speed of Yuri Gagarin and John Glenn was about 17,500 miles per hour—enough to circle the planet at their altitude in about 90 minutes. This contrasted with the record of 2,196 miles per hour previously reached by Joseph A. Walker in the X-15 on August 4, 1960. During orbital flight both Gagarin and Glenn soared higher than 160 miles. The advent of space flight now meant that man had to create a new time scale and drastically readjust his earthly concepts of speed and distance.

With the personification of space conquest in the form of dedicated astronauts, and with the inauguration of manned flights, our national and personal involvement in this latest frontier reached a new intensity. On May 25, 1961, President Kennedy underscored our commitment to space and made our objective precise and bold when he stated that the United States should achieve the goal, "before this decade is out, of landing a man on the moon and returning him safely to the earth. No single space project in this period will be more impressive to mankind, or more important for the long-range exploration of space. . . ."

Thus the target was established, and eventually 20,000 corporations, 300,000 people, 10 major NASA centers, and well over 23 billion dollars were dedicated to the task of reaching it.

Step by step through the space walks and intricate docking maneuvers of the two-man Project Gemini missions, and then the three-man flights of the Apollo series, the American people saw progress toward that target.

Finally, after eight years of preparation during which a tragic fire took the lives of three of the astronauts, the great Saturn V rocket stood poised, 36 stories tall, on pad 39A at the Cape. The date: July 16, 1969. The crew: Neil A. Armstrong, Edwin E. Aldrin, Jr., and Michael Collins. The specific objective: an elliptical landing zone on the moon's Sea of Tranquillity.

The historic launch came on schedule at 9:32 a.m., and the unprecedented odyssey of Apollo 11 had begun.

Four days later, the largest television audience in history (an estimated 500 million) saw the ghostly figure of Neil Armstrong clinging to *Eagle's* ladder as one foot cautiously tested the surface below. Then, as both feet came down solidly on the stark lunarscape, he uttered the now famous phrase: "That's one small step for a man, one giant leap for mankind."

The returning Apollo 11 crew brought back 47 pounds of lunar material. From these precious

FIRST AMERICAN TO ORBIT *the earth, Lt. Col. John H. Glenn, Jr., flashes battery-powered fingertip lights in a preflight test at Cape Canaveral. He used them during times of darkness on his 83,450-mile journey of February 20, 1962. Soon after the mission, Vice President Lyndon B. Johnson, on behalf of the National Geographic Society, presented the astronaut with the Hubbard Medal, the Society's highest honor in the field of research and exploration.*

NATIONAL AERONAUTICS AND SPACE ADMINISTRATION

SPACE AGE TOOLS AND TECHNOLOGY FIND DOWN-TO-EARTH USES

From a rocket's searing flame to instruments so sensitive the movement of an eye can activate them, from space towers to reduced-gravity simulators—man employs on earth the devices that have taken him far from his home planet. He applies his vast new technology in science and medicine—and in the improvement of his own earthbound environment.

UNITED TECHNOLOGY CENTER, SUNNYVALE, CALIFORNIA

COMPACT CUTTING TORCH *with a rocket-fuel flame frees occupants of smashed autos. Carried in patrol cars, the device may save thousands of trap victims who might otherwise die w waiting for emergency rescue vehic*

SPACE BLANKET, *made of lightweight insulation used in Apollo flights, reflects the sun's heat and shades Geographic staff photographer Dean Conger, at work in the Philippines.*

KENNETH MACLEISH, N.G.S. STAFF

NASA

HELMETED YOUNGSTER *takes an ear test, unaware of the electrodes touching his head. Doctors measure his hearing by watching how brain waves change when sounds come through earphones.*

NASA

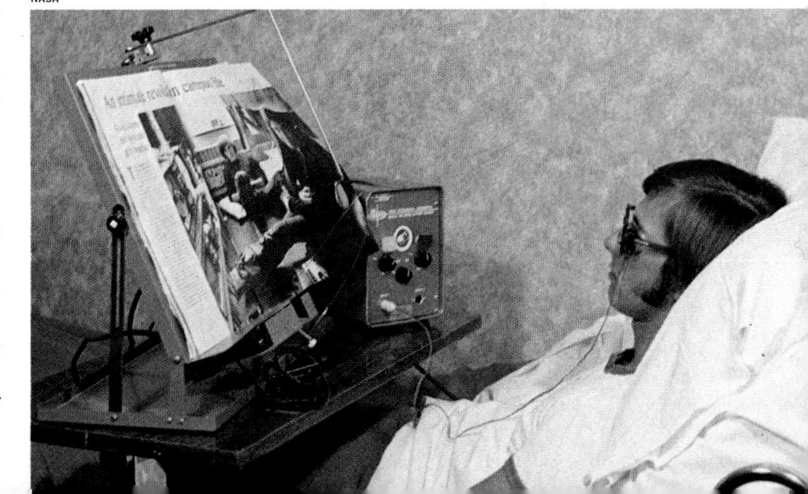

EYEBALL MOVEMENTS *control a device that turns pages for an immobile patient. Delicate light-beam switches that enabled astronauts to maneuver spacecraft simply by shifting their eyes can control TV sets, turn off lights, adjust beds in sickrooms.*

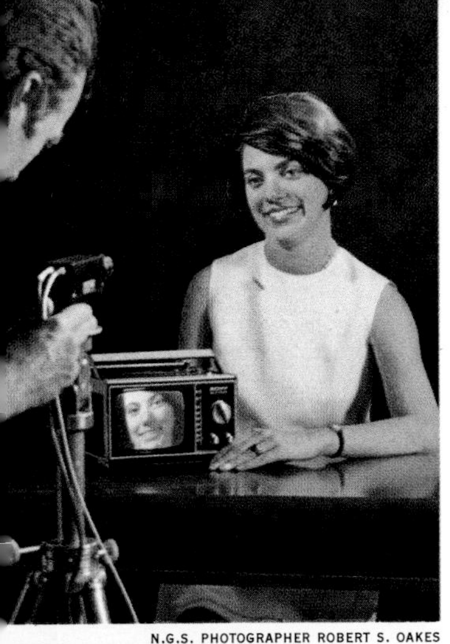

N.G.S. PHOTOGRAPHER ROBERT S. OAKES

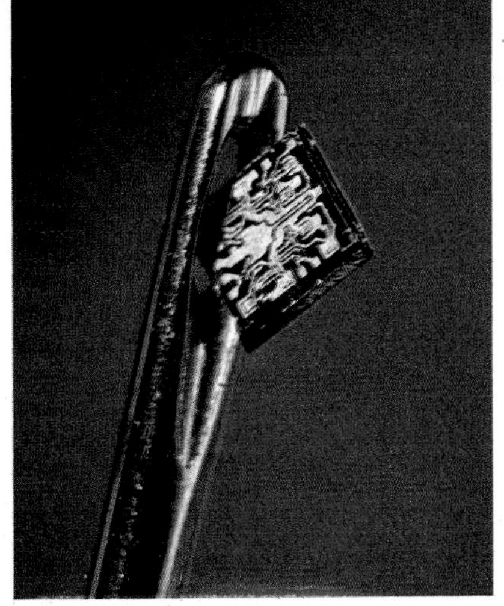

VICTOR R. BOSWELL, JR., N.G.S. STAFF

SILICON CHIP, *tightly packed with electronic circuits, can easily pass through the eye of a needle. Such tiny components, developed from missile technology, make possible a television camera barely larger than a package of cigarettes. Its uses on earth include underwater research and mobile news coverage.*

UNITED TECHNOLOGY CENTER, SUNNYVALE, CALIFORNIA

GLASS FIBER *for rocket-engine casings finds use in corrosion-resistant pipe, such as the length of water main being buried (left) at Grass Valley, California. The pipe, strong and resilient, weighs a sixth as much as reinforced concrete.*

TWEIGHT TOWER, *designed for space llations, stores in a barrel, expands)0 feet. Among earth uses: radio and her towers, utility poles, ladders.*

BELL TELEPHONE LABORATORIES

VICTIMS *of partial paralysis exercise in a reduced-gravity simulator devised to acclimate astronauts to space flight. With the help of his physical therapist, a young patient moves sideways at the Texas Institute for Rehabilitation and Research in Houston.*

A. Y. OWEN

NASA

BLUE CRAB *wears a one-inch-square transmitter that can send measurements of the creature's blood salinity through five feet of seawater. Such information may help ecologists determine just how pollution affects marine life.*

samples and others acquired on subsequent Apollo visits, man learned that his moon is a far from simple object, and its role as window to the past and door to the future is as complex as it is unique. Obviously comprehension of that role requires intricate and sustained analysis.

The Russians discovered this for themselves, not initially through exploration by men but by remote-controlled lunar vehicles. Lunokhod 1 landed on November 17, 1970, then spent several months in roving investigation. In September 1970, the unmanned Luna 16 brought back about 3.5 ounces of material from the Sea of Fertility, and in early 1972 Luna 20 returned samples from the Apollonius Mountains.

While the moon was presenting a host of tantalizing enigmas, the closest planets, Mars and Venus, were being investigated by unmanned spacecraft sent both by the United States and by the U.S.S.R.

Then on June 7, 1971, the Soviets achieved one of their most ambitious objectives. Three cosmonauts flying Soyuz 11 successfully docked with the space station Salyut, placed in earth orbit seven weeks earlier. The achievement marked the threshold of an entirely new dimension of space investigation.

But the first years of the Space Age had already brought manifold benefits to earthbound men in very practical ways. We watched events on other continents as they occurred, thanks to satellites that relayed TV pictures. Meteorological satellites warned us of hurricanes as they formed. Jet pilots routinely used satellite photographs of weather along over-water routes.

By 1971 the global satellite communication system's 10,000 circuits could handle more than 10,000,000 transoceanic calls annually, and the cost of a three-minute New York-to-London call had dropped from $9 to $5 in six years.

Remote-sensor satellites were effectively assisting in navigation, geodesy, cartography, geology, oceanography, mineralogy, and agriculture. (Infrared film, for example, reveals the spread of plant diseases so accurately that orchardists can determine which trees to spray; sensors can also measure the salinity and moisture content of soil.) In combination, such studies help forge solutions for a great range of the environmental problems of what Astronaut Frank Borman called our "small and beautiful and fragile" planet.

Scores of new industrial processes have evolved from the standards of precision, reliability, durability, and miniaturization dictated by the demanding conditions of space exploration. New families of plastics and alloys have appeared, including fabrics, foams, papers, paints, and fiberglass laminates resistant to heat, fire, corrosion, and cold. And the new management techniques and data processing methods developed in the space program show promise for application in many other fields of effort.

In medicine we have gleaned a host of benefits. Remote sensing devices used to monitor pulse, respiration, and other body functions of astronauts are increasingly employed in laboratories, hospitals, and ambulances. Special garments derived from astronauts' pressure suits have been adapted for cardiovascular patients and soldiers with abdominal wounds. New diagnostic, therapeutic, and prosthetic devices have been developed, including miniature probes that take pictures inside the body, artificial limbs and organs, even test models of a motorized wheelchair that severely crippled patients can control with eye movements. The same kind of computer that interprets photographic details of the moon makes possible a better and faster look at chromosomes and thus helps diagnose hereditary disorders.

By the early 1970's the new science of astronautics had embraced nearly every scientific discipline, from astrophysics to zoology, and it was apparent that significant portions of our human and technical resources were as permanently committed to space exploration as they had earlier been committed to the development of aviation. The mystique of space had secured a firm grip on the consciousness of man.

"The earth is the cradle of humanity," wrote the father of Russian rocketry, Konstantin Tsiolkovsky, "but mankind will not stay in the cradle forever."

CLOUDS WREATHE THE BLUE PLANET, *photographed at a distance of 19,000 miles from the explosion-crippled Apollo 13 spacecraft in April 1970. Clear skies over the southwestern United States and northwestern Mexico reveal the narrow peninsula of Baja California. Mission commander James A. Lovell, Jr., called our world a "grand oasis in the vastness of space," and observed, "We do not realize what we have on earth until we leave it."*

NASA

2/ SCIENCE AND FANTASY, A CHRONICLE OF SPACE

One clear night when I was picnicking with friends on a beach beside the Gulf of Mexico, one of the men suddenly pointed skyward. "There's a satellite," he yelled. A few seconds later his wife also pointed. "I see it," she exclaimed. "It looks just like a moving star!"

The rest of us were momentarily confused because they were pointing in different directions, but we quickly realized that not one but two artificial satellites were clearly visible overhead. Twenty minutes later we spotted a third pinpoint of light tracing its silent path across the dark sky.

These visible and conspicuous symbols of the Space Age still fill me with a sense of awe and mystery. Ancient man must have been no less awed by the lights he saw in the night sky. For him, the moon dominated the starry heavens, rising and setting and regularly changing its shape. At long intervals he beheld the intrusion of alien lights with long tapering tails. And he noticed numerous fiery streaks, ranging from the apparent size of embers to great bolts of light that exploded in soundless incandescence. From time to time the skies glowed with strange lucent clouds, pulsed with eerie shrouds and wraiths, or appeared to come alive with mysterious luminosities.

Of all objects in space, the daytime star, the sun, played the dominant legendary role. To the Egyptians, the sun was a god who had to be appeased by offerings from earth. The ancient Chinese believed a shaggy dwarf named P'an Ku created the sun—as well as the rest of the universe—with hammer and chisel.

In a more practical sense, ancient man used the sun to tell the time of day, and the time of a year composed of 365 sunrises. As early as 3000 B.C. the Egyptians and the Sumerians had devised and refined calendars. With the calendars, certain learned people, sometimes referred to as the natural astrologers, often predicted accurately the movements of heavenly bodies. This rudimentary ability to project human intelligence into the

(Continued on page 35)

EARTH'S AWESOME STAR, *an object of mystery and worship through the ages, bestows warmth and beauty upon mankind. Here the great yellow disk sets the horizon ablaze at Big Bend National Park in southern Texas, and spins an arc of light off the mirror of the photographer's reflex-camera lens.*

NATIONAL GEOGRAPHIC PHOTOGRAPHER JAMES STANFIELD

CAIRO MUSEUM, AUDRAIN-SAMIVEL

HAND-TIPPED RAYS *of the sun god Aten reach earthward to scatter the gift of life. In hollow relief, Egypt's Pharaoh Akhenaten and Queen Nefertiti present offerings at a religious ritual of the 14th century B.C. Condemning lesser deities, the king proclaimed one god only—Aten. Worshipers built luxurious temples and exalted the sun in hymns.*

MYTHICAL SPACE TRAVELER, *Icarus plunges to his death in the Aegean Sea after his wings of feathers and wax melted when he flew too close to the sun. His father Daedalus, who fashioned the wings, watches helplessly. Sketched in oil by the 17th-century Flemish painter Peter Paul Rubens, the Greek myth reflects man's ages-old longing to travel into space.*

ROYAL MUSEUM, BELGIUM

DETAIL OF PAINTING BY GIOVANNI DI PAOLO, COURTESY LEHMAN COLLECTION

EARTH-CENTERED UNIVERSE: *Paths of the planets appear as colored bands circling a fancied world in this detail of a 15th-century panel, "Expulsion of Adam and Eve From Paradise." Set forth by the astronomer Ptolemy in the second century, this concept prevailed for more than a thousand years.*

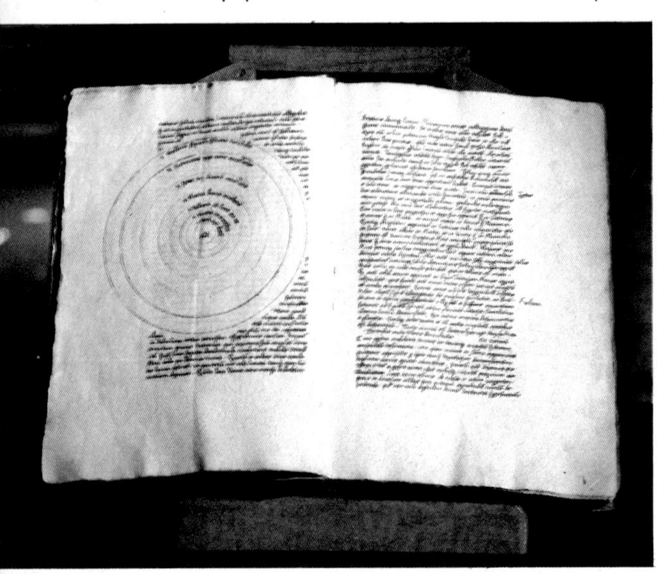

WOODEN QUADRANT *still gauges the sun's angle at Frombork, Poland, in the study of Nicolaus Copernicus, 15th-century astronomer who formulated a new theory of the universe, placing the sun at the center of the solar system. His treatise, at Krakow's Jagiellonian University, shows his concept of the orbits of the planets known in his day.*

ERICH LESSING, MAGNUM

FIRST SCIENTIST *to conduct a systematic study of the night sky through a telescope, Galileo Galilei mapped the stars and moon in 1609. His instrument (below, right) points toward the Doge's Palace from a balcony of St. Mark's Cathedral in Venice, Italy. The city's amazed Doge, Leonardo Dona, looked through Galileo's telescope at islands and ships in the Gulf of Venice. Before a window framing the leaning tower of Pisa rest two experimental spheres. Legend says that Galileo, intrigued by how objects moved, dropped such different-size stones from the tower and watched as they struck the ground almost simultaneously, thus disproving the theory that a large body would fall twice as fast as one half its weight.*

PAINTING BY GIUSTO SUSTERMAN, UFFIZI GALLERY, SCALA

ERICH LESSING, MAGNUM

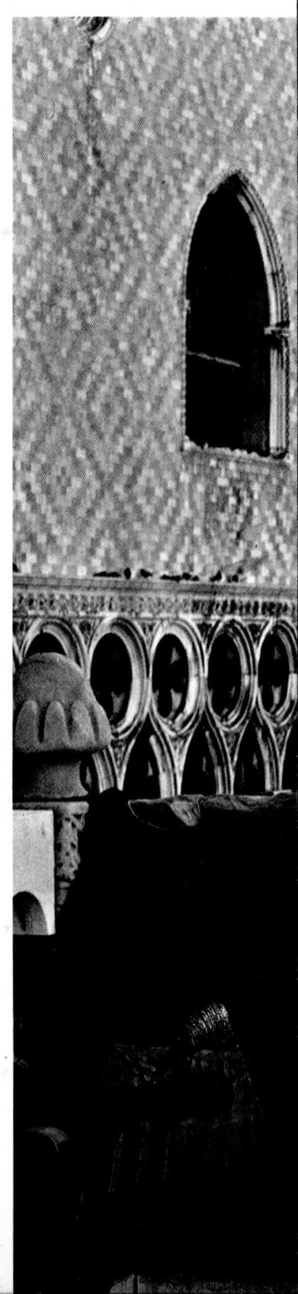

future encouraged another group, the judicial astrologers, to maintain that celestial objects could influence historical and natural events on earth. Thus, early study of the universe led to the belief that the fate of individuals and the advent of wars, floods, and plagues could be predicted by the relative position of heavenly bodies.

When the Greek astronomer Aristarchus first advanced the possibility—22 centuries ago—that the heavens did not revolve around the earth, few paid him any attention. Man preferred to believe that the earth was unique in the universe, and all else existed to serve it.

In the second century A.D. a Greek astronomer named Claudius Ptolemy, who lived in Egypt, published an encyclopedia of all man's beliefs concerning the universe. His *Almagest* accepted the view of another Greek, Hipparchus, that an immovable earth was at the center of the universe and that around it orbited the moon, the sun, and the planets. The Ptolemaic system remained supreme for more than a thousand years.

In 1543 the great Polish astronomer Nicolaus Copernicus published his theory that the planets revolved in perfect circles around the sun. Later, the brilliant German astronomer Johannes Kepler formulated new laws of planetary motion. Kepler's principal conclusion was that the planets orbit the

ENGLISH SCIENTIST *Isaac Newton, born in 1642, the year Galileo died, used precise mathematical equations to refine the Italian's theories. Whereas Galileo discovered how objects moved, Newton discovered why, explaining almost every motion in the universe with his Law of Universal Gravitation. On the windowsill of Newton's study at Woolsthorpe Manor stands a replica of his reflecting telescope. The instrument reflects light from a magnifying curved mirror near the base to a flat mirror near the eyepiece at the top. The desk holds a copy of Newton's* Mathematical Principles of Natural Philosophy, *his 1687 lexicon of mechanics, foundation of astronomy and physics.*

ERICH LESSING, MAGNUM

central sun not in perfect circles but in ellipses.

The telescope, as utilized by the Italian Galileo Galilei, abolished forever the concept of an earth-centered universe. Galileo, in the early 17th century, discovered four bodies that orbited Jupiter —not the resplendent central earth.

Half a century after Galileo's discovery, the Englishman Sir Isaac Newton used precise mathematics to state fundamental laws of motion and gravitation that applied equally to the fall of an apple and the movement of a planet. Orbital motion, said Newton, is the result of the balancing of two powerful forces: centrifugal force tending to cause a revolving body to continue on into space in a straight line, and gravitation tending to attract the revolving body.

Newton's later discovery that sunlight is the combination of many colors led to the spectroscope, through which astronomers were to discover not only the composition of heavenly bodies but also the approximate speeds at which they moved. In spectrographic portraits of distant galaxies, the displacement in their wavelengths, called the "red shift," tells us that the farther galaxies are away from earth, the faster they are receding from us.

The road from Isaac Newton to Albert Einstein was paved with new methods of mathematical

analysis and brilliant refinements in the discoveries and techniques of observational astronomy. In his Special Theory of Relativity, published in 1905, Einstein speculated that the velocity of light remains constant with respect to the motion of any body in the universe. His theory embraced a finite universe in which everything—matter and energy, space and time—is included and interrelated.

In regard to the universe as a whole, Einstein provided the basis for what came to be known as "the cosmological question." Did the universe begin with a "big bang," and is the universe still expanding? Did it evolve gradually, as stars and galaxies condensed from vast clouds of gas? Does it oscillate between periods of expansion and contraction? Or is it in a "steady state," without beginning or end, with new galaxies being formed as existing galaxies pass beyond the limits of the observable universe?

Some scientists looked for answers closer to home among the nine planets: Mercury, Venus, Earth, Mars, Jupiter, Saturn, Uranus, Neptune, and cold, dark, distant Pluto.

Until 1781 only the first six planets were known. In that year William Herschel of England discovered a seventh—Uranus. A few years later scientists applied Newton's law of gravitation to Uranus after they had noticed it deviated from its predicted orbit. Perhaps, they reasoned, Uranus was affected by the gravitational pull of an unknown planet. Astronomers determined the position of the mysterious force that caused the erratic behavior of Uranus, and in 1846 discovered the eighth planet, Neptune.

Neptune, too, showed fascinating orbital deviations. One day in 1930, a young assistant named Clyde Tombaugh was working in the Lowell Observatory among the ponderosa pines of Flagstaff, Arizona. Hunched over a device used for comparing photographs, he carefully studied pictures of the skies taken several nights apart. His purpose was to see if any one of millions of specks of light had changed position. Such an occurrence would mean that he had spotted "Planet X," then the object of a worldwide search.

Suddenly Tombaugh leaned forward. A single dot appeared to have moved. He quickly examined other plates. The tiny speck seemed to be traveling through space. Heart pounding, Tombaugh strode down the hall and entered the office of his superior, Dr. V. M. Slipher. Two years earlier Dr. Slipher had been so impressed with drawings of Mars Tombaugh had made with the help of his homemade telescope that he sent a letter to the Kansas farm boy and offered him a job.

"Dr. Slipher," Tombaugh announced, "I've found Planet X." The Kansas farm boy had indeed discovered the ninth planet—to which the Royal Astronomical Society gave the name Pluto.

I once talked with Clyde Tombaugh after he moved to Las Cruces to teach at New Mexico State University. "For 14 years," he told me, "I searched for a possible tenth planet, but I finally concluded that Pluto is our last one."

Tombaugh's preoccupation with celestial realms has never slackened, however. During the 1950's he originated a technique of searching near space for any small natural satellites hurtling around the earth that might prove hazardous to moon voyagers. His group examined 15,500 photographs and meticulously studied the sky without finding even one such object. In 1958 Tombaugh began a continuing study of planetary markings that has greatly influenced space probes. He and his co-workers have pored over photographic plates containing one million planet images. On the basis of this study, Tombaugh suggested areas on Mars for exploration by Mariner missions in 1969.

At about the same time we were able to comprehend a solar system of nine planets, a new, little-known branch of science—that of rocketry—began to emerge, stimulated in part by science-fiction accounts through the years.

Fictional journeys into space, though often depending upon such unorthodox means of locomotion as geese, swans, demons, and chariots, employed rocket principles as early as the 17th century. In *Voyages to the Moon and the Sun,* Cyrano de Bergerac introduced a machine powered by rockets. Igniting in quick succession, the rockets boosted him "into the clouds."

Ironically, the great astronomer and mathematician Kepler, when he turned from space science to space fiction, used highly unscientific methods. He solved the problem of propulsion by having

NATIONAL MARITIME MUSEUM, GREENWICH, ENGLAND

MUSIC MASTER AND ASTRONOMER, *William Herschel discovered Uranus in 1781 while studying the sky as a hobby. The German-born immigrant to England named his find the Georgian Planet after King George III; later he became the court astronomer.*

BOATLOAD *of gaily dressed space travelers returns from the moon by parachute, bringing captured lunar creatures and weird plants. This 19th-century print reflects the belief of Herschel and others of his day that life existed on the moon.*

1836 LITHOGRAPH BY GATTI AND DURA, LIBRARY OF CONGRESS

demons magically transport human beings to the moon during times of eclipse.

The Frenchman Jules Verne, whose prophetic writings in the 1860's influenced pioneer rocket scientists of the 20th century, developed the concept of steering-rockets to maneuver his spaceship in *From the Earth to the Moon* and in *Around the Moon*. The ship's initial impetus was delivered by a 900-foot-long cannon buried upright in Florida sands. However, the rapid acceleration necessary to propel the rocket into space would have killed the three-man crew instantly.

Around the beginning of the 20th century, H. G. Wells conceived a system of propulsion that was simplicity itself. In *The First Men in the Moon,* his spacemen entered a sphere enameled with a mysterious gravity-defying substance called Cavorite, opened a window aimed at the moon, and sped on their way.

Wells devised the idea of Cavorite at about the time it became clear that the Chinese, who had invented gunpowder and rockets centuries earlier, were on the right track all along; rocket propulsion was the practical means of sending into space cargoes of explosives, instruments, and human life itself.

The basic principle applied by the Chinese to launch their rockets has been observed by children for generations: A toy balloon expelling air provides an example of rocket propulsion. So does a firecracker that misfires and fizzles across the sidewalk, spewing smoke.

The Chinese and Mongols improved multiple arrows of "flying fire." Propelled by rockets, the missiles rushed out "on a solid front like 100 tigers." The Mongols carried the new rocket weapons to the Near East and ultimately to Europe. By the 16th century many Europeans had become fascinated with the "casing that flies" and with its potential in fireworks displays and in warfare. In 1668 a German field artillery colonel, Christoph Friedrich von Geissler, developed and fired several experimental rockets weighing up to 132 pounds.

Progress in military rocketry was rapid in 19th-century Europe. More than 25,000 British rockets developed by Col. William Congreve were launched against Copenhagen in 1807. The same type of rocket, weighing about 30 pounds, was used in the War of 1812. Francis Scott Key's words "the rockets' red glare" in "The Star-Spangled Banner" refer to the British rockets that bombarded Baltimore's Fort McHenry.

In World War I, rockets were used chiefly to launch signal and parachute flares that illuminated enemy positions at night. The French made limited use of rockets, firing them from Nieuport biplanes at barrage balloons and from the ground at German zeppelins. But, generally speaking, military rockets of the period were made nearly obsolete by improved artillery, the airplane, and radio—which in England saw experimental use as a means of directing a pilotless plane or "flying bomb," the precursor of winged missiles.

If rocket applications were somewhat limited, rocket theory was slowly emerging with a few isolated, highly imaginative individuals who were to influence profoundly the coming age of space. The first of these was a Russian, Konstantin Tsiolkovsky, born at Izhevsk on September 5, 1857.

On a trip to the Soviet Union in 1966, I asked authorities if I could visit Kaluga, where Tsiolkovsky worked in his later years. In a little Volga car, I drove to the railroad and machinery city on the banks of the Oka River some 125 miles from Moscow. Tsiolkovsky's grandson, Alexei Kostin, showed me the tools, models, and papers of the distinguished space scientist.

At the age of nine, according to Kostin, an attack of scarlet fever severely impaired Tsiolkovsky's hearing. Virtually shut off from the world of sound, his mind turned inward and dreamed.

"In my imagination," Tsiolkovsky wrote of his childhood, "I could jump higher than anybody else, climbed poles like a cat and walked ropes. I dreamed . . . there was no such thing as gravity."

After educating himself in mathematics and physics, Tsiolkovsky experimented with gravity-defying contraptions ranging from a "mechanical hawk" to balloons and an all-metal dirigible. Barely supporting himself on a teacher's stipend, he worked with increasing attention on the theory of

ALBERT EINSTEIN *revolutionized man's concepts of space, time, matter, energy, and light half a century before the first orbital flight. At age 26, he stated in his theory of relativity that all of these are not independent of one another, but complexly interrelated. The physicist's famous equation,* $E=mc^2$ *(energy equals mass times the speed of light squared), led to the production of nuclear energy, which may someday power space ships.*

NEW YORK TIMES

FIRST RECORDED APPLICATION *of jet propulsion: In this artist's conception, a wooden dove, possibly driven by steam or compressed air, fascinates Greek nobles. Archytas of Tarentum, who fashioned such a device around 360 B.C., stands near weights and strings that suspend the bird and keep it in equilibrium. Today's jets and rockets employ basically the same ancient principle of thrust.*

space flight. Stimulated by the fantasies of Jules Verne, Tsiolkovsky wrote:

"To place one's feet on the soil of asteroids, to lift a stone from the moon with your hand, to construct moving stations in ether space . . . to observe Mars at the distance of several tens of miles, to descend to its satellites or even to its own surface—what could be more insane! However, only at such a time when reactive devices are applied, will a new great era begin in astronomy: the era of more intensive study of the heavens."

In 1903 he published in the Russian *Scientific Review* his first article on rocketry, a theoretically sound primer on the liquid-propellant rocket such as the models I saw in his workshop. Later, he improved his drawings and explanations of rockets, of the use of kerosene as a fuel, of methods of cooling the combustion chamber, of various means of guidance, and of the multistage rocket, or what he called a rocket train.

By the age of 78, he had written scores of scientific papers and a number of science-fiction novels, but he still devoted half of his time, as he put it, "to the problem of overcoming terrestrial gravity and making flights into space."

Two years before his death in 1935, he made a prophetic declaration on a radio broadcast: ". . . I am firmly convinced that my . . . dream—space travel . . . will be realized . . . I believe that many of you will be witnesses of the first journey beyond the atmosphere. . . ."

The father of modern rocketry in America, Robert H. Goddard, was born on October 5, 1882, in Worcester, Massachusetts. As early as high school, he was writing papers on space, and by the time he received his doctorate from Clark University in Worcester in 1911, he had already discovered, as Tsiolkovsky did, that liquid oxygen and liquid hydrogen would make an ideal rocket propellant. Within four years, Goddard had carried out several successful field experiments, including the launch of powder rockets that rose as high as 486 feet above the Massachusetts countryside.

In 1917 the Smithsonian Institution granted him $5,000 to conduct high-altitude experiments. Two years later, a paper on his proposal, "A Method of Reaching Extreme Altitudes," was published, presenting a type of propulsion that theoretically could send a rocket to the moon. Goddard incidentally included a method of determining whether the rocket had reached its target: On impact, a mass of flash powder attached to the missile would explode and could be observed from earth by telescope.

The relatively insignificant flash-powder idea, not the rocket theory, caught the public fancy.

GEORGE FRANCUCH, FROM AC ELECTRONICS

The quiet mustached professor found himself the object of unwelcome attention. Thereafter shunning publicity, he alternated between his own rocket research and work on depth charges and armor-piercing rocket projectiles for the U. S. Navy in the early 1920's.

In March of 1926, a 12-foot projectile, the world's first liquid-propellant rocket, rose from the farm of Robert Goddard's Aunt Effie near Auburn, Massachusetts. The short flight, before the engine malfunctioned, was reminiscent of Kitty Hawk: height 41 feet; distance 184 feet; speed 60 miles per hour. Goddard sent up three more rockets at Auburn. The last, on July 17, 1929, was mistaken by residents for an airplane in flames, and the incident caused a flood of publicity.

From some of the newspaper accounts, Col. Charles A. Lindbergh learned of Goddard and his work and paid him a surprise visit in November. The great aviator, like Goddard, envisioned an expanding future of rocket travel and, ultimately, space flight. Lindbergh encouraged philanthropist Daniel Guggenheim to grant Goddard $100,000.

With four assistants, Goddard set up a simple launch facility near Roswell, New Mexico. There, in 1935, he launched a remarkably sophisticated 85-pound projectile. Its liquid oxygen and gasoline were fed into a combustion chamber from a

pressurized tank; the chamber itself was prevented from melting by the flow of propellants within its hollow casing—a procedure called "curtain cooling"; a gyroscope provided stability. The rocket rose 7,500 feet above the New Mexico desert. Another reached almost supersonic speed.

With continued financial support from The Daniel and Florence Guggenheim Foundation, Goddard built and flew dozens of intricate rockets during more than a decade of work in New Mexico. However, he was far better known abroad than in his own country. In Germany scientists used many of the same engineering principles as Goddard in developing the V-2 rocket used against Allied targets during World War II.

Germany was not without its own pioneer in the theory of rocket flight. The early calculations and writings of Hermann Oberth certainly must rank him along with Konstantin Tsiolkovsky as a major rocket theoretician.

The future professor of mathematics and physics was born on June 25, 1894, at Hermannstadt, Transylvania. "At the age of eleven," Oberth reveals in his autobiography, "I received from my mother as a gift the famous books . . . by Jules Verne, which I read at least five or six times and, finally, knew by heart."

While in his 20's, during World War I service in

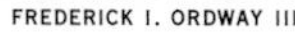
FREDERICK I. ORDWAY III

GUNPOWDER ROCKETS, *used by the British against Napoleon, arch above the battlefield at Waterloo on June 18, 1815, as a charging column of French horsemen attacks unwavering ranks of British foot soldiers. Rockets found use as weapons of war as early as the 13th century in both Europe and the Orient. At left, Sir William Congreve, developer of war rockets for Britain, watches Copenhagen go up in flames after a fierce incendiary barrage in 1807. Congreve's rockets saw their most celebrated moment in the British bombardment of Baltimore's Fort McHenry on the night of September 13, 1814, during the War of 1812, when Francis Scott Key immortalized "the rockets' red glare" in verse.*

the Austro-Hungarian Army, Oberth unsuccessfully proposed liquid-propellant long-range bombardment missiles to the German War Ministry. Later, after learning of Goddard's work from a newspaper account, Oberth corresponded with him, suggesting that they exchange their writings. "I think that only by the common work of scholars of all nations," Oberth wrote, "can be solved this great problem . . . to pass over the atmosphere of our earth by means of a rocket."

In 1923, Oberth published a thin but highly influential book called *The Rocket into Planetary Space,* in which he provided an enlightened and practical design for a complex liquid-propellant rocket to explore the upper atmosphere. Six years later, he published a 423-page expansion, *The Road To Space Travel.* The same year he became president of Germany's Society for Space Travel.

Oberth, hoping to stimulate public interest in his field, consented to launch a rocket for a German movie. The attempt failed but attracted widespread attention, especially from Germany's growing roster of young rocket enthusiasts. The German-born author Dieter K. Huzel wrote of this period, "As model airplanes captured the enthusiasm of American youngsters in the 1930's, so rockets were a source of endless excitement, and an even more challenging toy, for German children of the 1920's."

In 1932, the Society for Space Travel moved to the Kummersdorf proving ground near Berlin to test rockets. Among early members of the group were a handsome, firm-jawed student named Wernher von Braun and a young engineer, Walter Dornberger, who persuaded the German army to donate a modest sum to finance test flights. By 1934, eight years after Goddard fired his first liquid-fueled rocket, the Germans got two such projectiles to an altitude of 6,500 feet. By 1937, despite the financial depression in Germany, the

SHUTTLING PASSENGERS *to the moon, a coal-burning missile (opposite) speeds through space on a fanciful voyage inspired by the writings of Frenchman Jules Verne. In his 1865 classic,* From the Earth to the Moon, *a "spout of fire" from a buried cannon launched a space capsule. In a sequel,* Around the Moon, *Verne anticipated the effects of weightlessness on space travelers. The author, amazingly prophetic, had his moon ship zoom skyward only 120 miles west of present-day Cape Kennedy.*

LIBRARY OF CONGRESS

FIRST CLASS
FIRST CLASS

Dornberger-Von Braun group numbered nearly a hundred and had expanded activities to a new test site on the Baltic coast called Peenemünde.

In Russia, by the early 1930's, government-backed scientists in Leningrad and Moscow were already gaining team experience in building and testing solid- and liquid-fueled rocket motors. By the mid-1930's, Russian rockets were reaching altitudes up to 3.5 miles. Around 1940, the Russians developed a rocket with a 12-mile range, using a combination of solid and liquid propellants.

By then all three countries had suffered temporary setbacks. Goddard postponed his work to assist in military research projects. In Moscow's sensitive political atmosphere, expanding Soviet rocket societies became suspect. And Germany's capricious chancellor, Adolf Hitler, dreamed one night that missiles wouldn't reach England, and for a time withheld development funds. But despite the shortage of money, the German army continued its work on rockets.

World War II brought about a resurgence of rockets of all types, surface-to-air, air-to-air, and air-to-surface. These included batteries of comparatively small solid-propellant rockets which the German and Russian armies fired at each other in barrage fashion.

The United States employed a number of rocket arms; one of the best known was the foot soldier's bazooka, a shoulder-held launcher inspired by Goddard. The air forces of all major combatants developed both jet and rocket engines to give an added boost to heavily loaded planes at takeoff.

But it was the supersecret products that rose

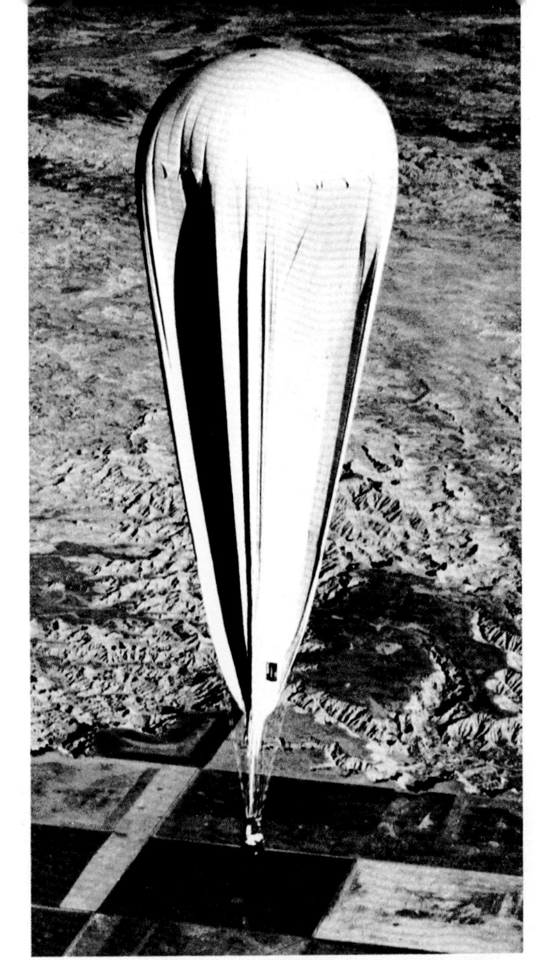

M/SGT. G. B. GILBERT AND CAPT. H. K. BAISLEY (ABOVE) AND THE BELL FAMILY

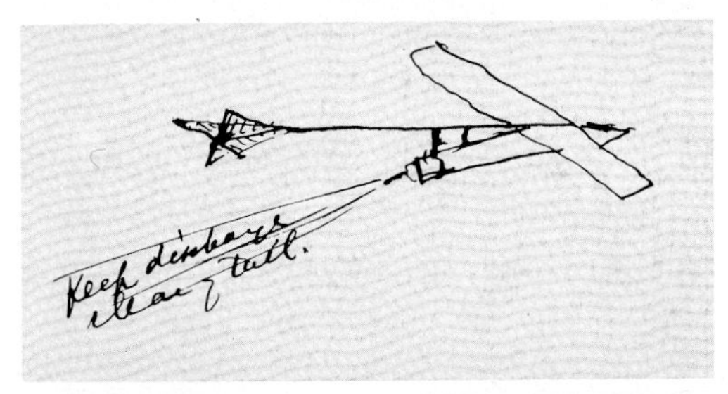

LIFTING MAN *to the stratosphere,* Explorer II *rises above South Dakota on a mission sponsored in 1935 by the National Geographic Society and the U. S. Army Air Corps. The balloon reached an altitude of 13.71 miles—a record that stood for 21 years. Decades earlier, in 1893, Dr. Alexander Graham Bell (right), experimenting with new propellants, built a rocket-powered model aircraft that flew 75 feet. The notation on his sketch warns against blast that might buckle the thin metal tail.*

RUSSIAN THEORIST *Konstantin Tsiolkovsky (left), first to suggest the use of rockets for space travel, studies at his home in Kaluga. In the early 1900's Tsiolkovsky advanced the idea that powerful rockets could escape the earth's gravitational pull.*

NOVOSTI

COURTESY ESTHER C. GODDARD (LEFT)

NATIONAL GEOGRAPHIC PHOTOGRAPHER B. ANTHONY STEWART (ABOVE AND RIGHT)

PIONEER OF MODERN ROCKETRY, *American Robert H. Goddard keeps a finger on the ignition key as he sights the launch tower at Eden Valley, New Mexico, in 1940. Combining theory and practice, Goddard created the first rockets to use liquid oxygen and gasoline as fuel. Crewmen at left prepare a rocket for firing. At right, Goddard inspects one of his test projectiles on its assembly frame.*

from the Peenemünde development pads that gave rocketry its most astonishing thrust. From 1938 on, the Dornberger-Von Braun group tested a succession of liquid-fueled rockets. In 1943, Hitler, in his desperate search for *wunder-waffen*—wonder weapons—finally decided to give top priority to the gasoline-burning V-1 "buzz bomb," and the 46-foot V-2 ballistic rocket.

By 1944, the first of more than 4,000 V-2's were launched. With a speed of 3,600 miles per hour, a range of up to 200 miles, and a ton of explosives, the V-2 quickly proved it was a true "wonder weapon" against which there was no known defense. Flaming alcohol and liquid oxygen pushed it skyward. Once its engine shut down, it arched through the edge of space to its target. Its deadly payload then plunged downward much too fast for interception.

The V-2, combining the best concepts of Tsiolkovsky, Goddard, and Oberth, represented by far

21 22 23 24 25 26 27
28 29 30

the most brilliant single forward stride thus far taken in rocketry.

Although born as a weapon of war, the V-2 represented much more than that to the rocket scientists; it was an exciting new means of transportation, like the steamship, the railroad, and the airplane, whose future cargo could include everything from mail to human beings. And, most important, it alone of all means of transportation was unconfined by the atmosphere of earth.

At war's end, Peenemünde held the most significant war booty of modern times. But when the Russians reached the site, they found a few rockets and little else. Unknown to them, Von Braun and his colleagues had voted to turn over their knowledge to the Americans. As the heart of the Peenemünde technical team moved south before the advancing Russians, they led a curious convoy of trucks and trailers. Inside the trucks reposed the key plans and blueprints not only of the V-2 but of even larger rockets capable of transatlantic flight.

Dieter Huzel, who cautiously shepherded the precious trucks, later wrote in his book *Peenemünde to Canaveral,* "These documents were of inestimable value. Whoever inherited them would be able to start in rocketry at that point at which we had left off, with the benefit not only of our accomplishments, but of our mistakes as well—the real ingredient of experience. They represented years of intensive effort in a brand-new technology, one which, all of us were still convinced, would play a profound role in the future course of human events."

GERMAN V-2 ROCKETS *crowd a production depot at Nordhausen in World War II. Riding a narrow-gauge track, a missile passes beneath a no-smoking sign, bound for one of scores of launching sites. More than 4,000 of the projectiles, perfected late in the war at Peenemünde, roared toward Allied targets at 3,600 miles an hour. German pioneer Hermann Oberth (left), who believed rockets could launch vehicles into space, inspired experimentation that led to the V-2. Below, 18-year-old Wernher von Braun, a student of Oberth's, carries an experimental rocket at a Berlin testing ground in 1930.*

DER SPIEGEL (BELOW) AND FREDERICK I. ORDWAY III

37
11/W4156
11/W4171
Rauchen verboten.

3/ BEYOND THE WALL OF AIF DANGERS OF THE COSMO

Astronaut Charles Conrad, Jr., and I had obviously not been alone in our decision to go for a Sunday-afternoon skyride. As we swooped toward the airfield near Houston's Manned Spacecraft Center, a swarm of small planes seemed to fill the air, darting like bees above a field of clover. Shoving the throttle forward, "Pete" Conrad shook his head and quickly nosed the light plane toward less-crowded skies. "Give me space!" he shouted. "I'll take space any day!"

He could make the comparison from extensive experience: Three times Pete has ridden a rocket into the sky, and he has spent a total of more than 500 hours in space. On two of those flights he helped set records of sorts: for altitude attained, and for sheer enthusiastic response to space adventure. After the 853-mile height he and Astronaut Richard F. Gordon, Jr., reached in Gemini 11 in September 1966, he exclaimed, "This has got to be the greatest ride in the world!" Three years later, he, Gordon, and Alan L. Bean traveled 228,216 miles to the moon during the Apollo 12 mission. Many people still remember the exuberance of the irrepressible Princeton graduate almost a quarter of a million miles away: "Whoop-de-dee! . . . Heigh ho! Heigh ho! Up the ladder we go!" During his lunar prospecting, some recognized the tune that came through the gap between two front teeth: "Whistle While You Work."

The glee with which Conrad explored space in no way detracted from the scientific accomplishments of his flights. He was as aware as anyone that on his Gemini 11 flight, he and Gordon were doing what scientists over the centuries have wished to do: observe the earth from afar and, without the interference of the atmosphere, study the planets, the sun, and the stars.

Within five minutes after lift-off by the powerful Titan II rocket, Conrad and Gordon were in earth orbit 165 miles high. They had quickly left behind the flesh-distorting g-forces created by rapid acceleration, and were riding weightlessly

DISCOVERER *of Pluto, Clyde W. Tombaugh gazes through the 24-inch telescope at New Mexico State University in Las Cruces. He spotted our outermost planet in 1930 while studying night-sky photographs. Astronomers, probing with Space Age telescopes, cameras, and satellites, add increasingly to man's knowledge of the universe.*

N.G.S. PHOTOGRAPHER EMORY KRISTOF

NORTHERN LIGHTS, *caused by the interaction of charged particles from the sun with atoms and molecules in the upper atmosphere, undulate above Fairbanks, Alaska. Called aurora borealis in the Northern Hemisphere and aurora australis in the Southern, they normally occur between 70 and 250 miles high and can last for several hours.*

DR. SYUN-ICHI AKASOFU

around the earth at 17,500 miles per hour. Their flight speed maintained an outward force equal to the inward pull of gravity, holding them in an orbital track. If they had stepped out of their spacecraft untethered, they too would have become orbiting objects whose course Newton could have described with an equation.

When the astronauts came back to earth after three days, they had felt and seen proof of the physical laws of space in operation. Closer to the stars than any man before him, Gordon had hung out the open hatch of the craft, taking pictures of celestial bodies, photographing our own sun as it poured energy on a turning earth.

How easily we can forget, down here under our protective blanket of air, that our sun is an immense nuclear reactor! Every second it transforms some four million tons of its matter into electromagnetic energy that streams into space. The rays travel at 671 million miles an hour—186,000 miles per second—across space; particles of atoms, boiling off the sun, move more slowly and create a solar wind that sweeps past the planets.

Only about one two-billionths of the sun's radiation reaches us, but if we had no protection, especially during great outbursts, or flares, on the sun, the rays and fragments would create a sizzling, deadly environment on earth. Fortunately, two lines of defense surround the earth—a magnetic field and the atmosphere.

The sun's rays—visible light, invisible infrared, ultraviolet, radio waves, and X-rays—reach us first. They take only eight minutes to speed the 93 million miles to earth. The slower solar wind protons (hydrogen nuclei), electrons, and scraps of other elements normally arrive in about four days, but during a flare they may come in 8 hours.

Surrounding earth, the magnetic field holds off or interrupts many of the particles. The invisible force originates at the earth's magnetic core, believed to be a molten mass of nickel-iron. Sweeping out from earth in majestic north-south arcs, the field reaches 40,000 miles into space.

"The deflection of solar wind particles," says physicist Donald J. Williams, formerly of Goddard Space Flight Center, "sets up mechanisms, not yet fully understood, which are responsible for the now familiar Van Allen radiation belt. The band encircles the earth opposite the middle latitudes where most people live. No one knew of the belt until Geiger counters on the first U. S. unmanned satellites, Explorer 1 and 3, revealed concentrations of amazingly high radiation counts."

Conrad and Gordon penetrated the lower fringe of the region. Instruments on board measured its radiation and confirmed earlier indications by

satellites that men inside a craft passing quickly through it would not be harmed.

A few particles are lost from the trap and rush into our atmosphere. They collide with air molecules, lose speed, and become part of the atmosphere. In the polar regions, where the Van Allen belt does not reach, great quantities of the particles from distant regions of the magnetosphere may fly in along the outer limits of the magnetic field. Pouring into the upper atmosphere near the poles, they collide with air molecules and thus produce shimmering auroras.

How are we protected from the solar rays—ultraviolet and X-rays that have short, extremely energetic wavelengths? They pass through our magnetic field unhindered, for it can entrap only electrically charged particles. But about 40 miles from earth they are blocked by a rapidly thickening zone of our atmosphere. Those with the shortest wavelengths use up their energy as they shatter the air molecules—and chip electrons from atoms, changing the atoms into ions, or positively charged particles. The region of ions and free electrons, called the ionosphere, not only protects man from solar radiation but also serves as a reflector for long-distance radio waves.

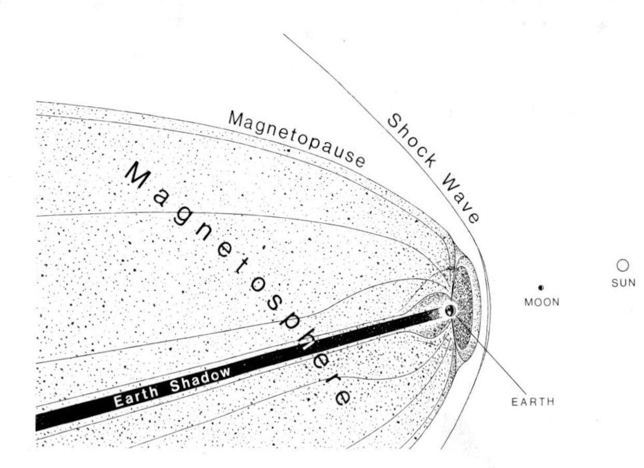

EARTH'S VAST, ENVELOPING MAGNETOSPHERE *(shown in cross section and indicated by purple and green areas) filters solar and galactic radiation. This gigantic envelope results from the collision of earth's magnetic field with the solar wind—a continuous flow of electrons and protons emitted by the sun. The boundary of the magnetosphere, the magnetopause, occurs where the pressure of the solar wind equals that of the magnetic field. Compressing the field to within about 40,000 miles of earth on the side facing the sun, the solar wind flows along the*

magnetopause and defines the shape of the magnetic field, believed to emanate from earth's molten nickel-iron core. The part of this field originating in the polar regions is pushed away from the sun by the wind and trails out two million miles or more, forming the magnetic tail. At the unprotected polar caps, solar and galactic cosmic rays (yellow and orange streaks) easily penetrate to the atmosphere. The collision of the solar wind with the field produces a shock wave (bowed blue line) 15,000 miles in front of the magnetopause in the direction of the sun. Trapped charged particles encircle the earth, forming the Van Allen radiation region (green areas). It extends 30,000 miles beyond earth; the heaviest concentration of protons and electrons occurs at 1,500 and 15,000 miles. Low-energy particles (purple area) move through the rest of the magnetosphere. The moon, lacking a strong magnetic field or heavy atmosphere, gets full exposure to solar radiation on its sunward side, except when sheltered by earth's magnetic field for about four days around the full moon.

Still the danger from rays is not completely over. Below the ionosphere, at an altitude of about 20 miles, another crucial defender waits: Molecules of ozone, a form of oxygen, absorb the energy of the longer ultraviolet rays. Finally the stratosphere and the dense troposphere, which holds the air we breathe, soak up most of the energy of other rays and particles, letting through only life-giving light, benign radio waves, some infrared, and just enough ultraviolet radiation to tan—or burn—the skin. At last we are safe, even from the awesome barrage of solar-flare radiation.

High above in space, Conrad and Gordon in their craft and suits are safe, too—except during a flare. Dangerous eruptions usually come weeks or months apart in more active periods on the sun, and occur rarely at other times.

While minute doses of high-energy protons, electrons, and X-rays over the whole body during a short time have no noticeable effect, heavy doses may cause vomiting within two hours. If enclosed only in his space suit, the astronaut could inhale the weightless, floating material and suffocate. If he escaped this danger, he might later develop some form of cancer from the radiation.

To avoid these threats, ground control would terminate the mission in the event of a flare.

No unexpected solar flare troubled the flight of

VICTOR R. BOSWELL, JR., N.G.S. STAFF (BELOW); WILLIAM BELKNAP, JR.

METEORITICIST *saws through a one-ton boulder from space. H. H. Nininger, outside his home at Sedona, Arizona, worked 162 hours to reveal for study a cross section of this meteorite, brought to the U. S. from the Philippines. Symbols in white tape mark its many magnetic poles—most meteorites have a single positive and a single negative pole. Some scientists believe that glassy tektites (above) splashed to earth a million years ago when a huge meteorite struck the moon and formed the crater Tycho, and that the friction of the impact and of their fall through the atmosphere heated and shaped the fragments.*

Gemini 11. Conrad, gazing out into the velvet-black backdrop of space, of course could see nothing of the invisible, normal turbulence going on in the magnetosphere and atmosphere as rays, electrically charged particles, atoms, and molecules interacted. He saw instead smoke and smog blanketing parts of the earth, some brush fires in Africa, clouds and lightning, and the airglow—an emission of faint green light from bands of oxygen atoms and a yellow light from bands of sodium atoms about 60 miles above the earth.

The astronauts took pictures of them all. They took pictures, too, of the sun, unobscured by our atmosphere as night changed into day along a definite line on the rotating earth. Like a distant fire in the night, the sun's own disk is bright while the expanse of space around it remains black, except where an object reflects the light—such as earth; Gemini and its rocket; or even occasional specks of matter darting past like fireflies.

Perhaps the oddest aspect of space travel is weightlessness. Conrad, tied to his seat, watched with fascination as "all the debris and junk . . . in the spacecraft went out the window" when Gordon opened it for his walk. "And I was right along with the rest of the debris," Dick said.

Floating idly was restful, but floating as he tried to work made Gordon sweat and gasp, and his

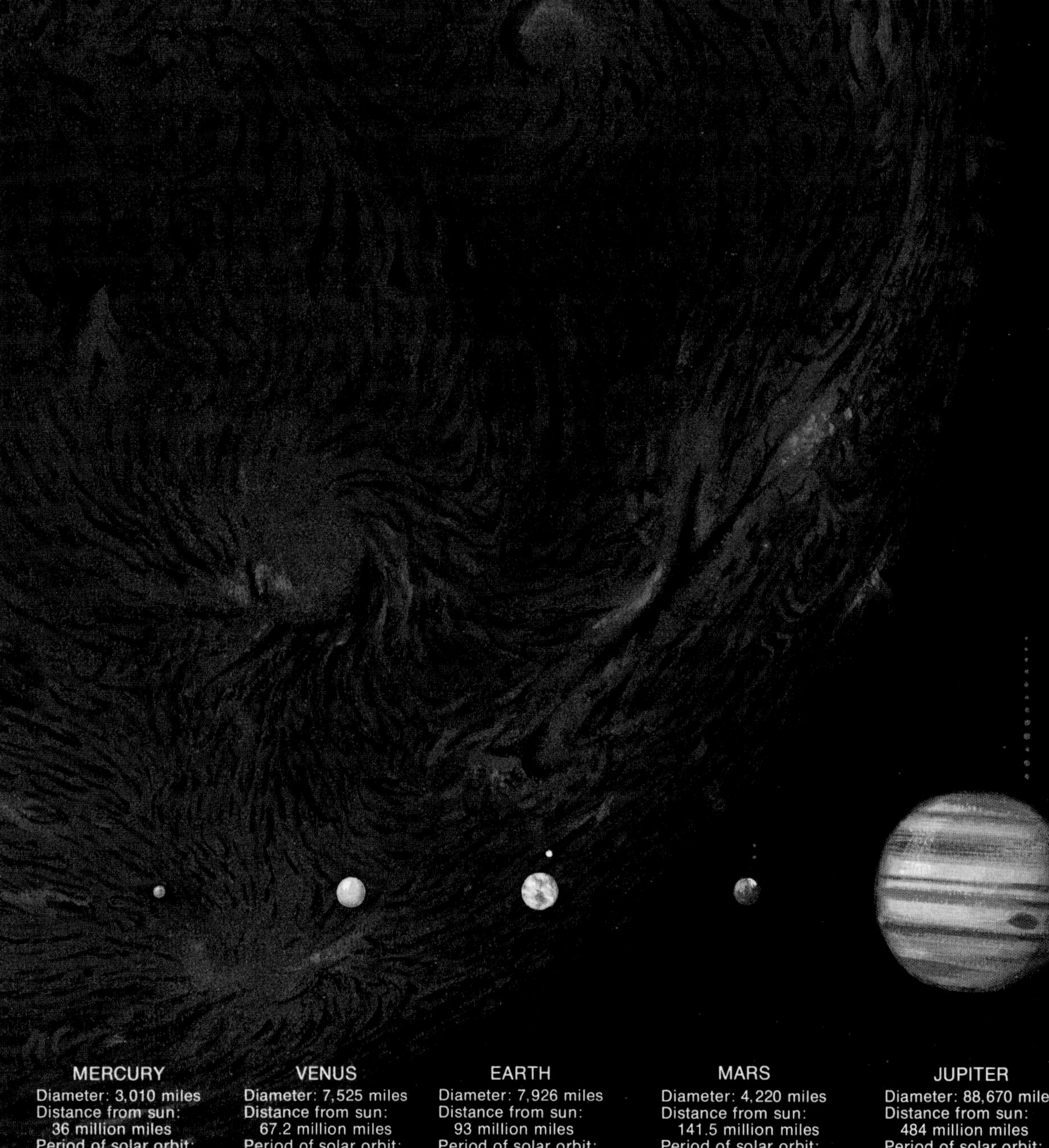

DOMINATING OUR SOLAR SYSTEM, *the sun holds nine planets in orbit with its powerful gravitational force. This typical star, a nuclear furnace nearly 900,000 miles in diameter, contains 99.86 percent of the solar system's matter. Planets—from the Greek word for wanderers—divide into two groups according to their properties. Mercury, Venus, Earth, Mars, and possibly distant Pluto, make up the terrestrial planets, composed of nickel-iron and a mixture of rocky materials. Hydrogen and helium form the bulk of the larger and more massive Jovian planets—Jupiter, Saturn, Uranus, and Neptune. The orbital velocities of the planets vary greatly with respect to their distances from the sun: Tiny Mercury hurtles around its stellar neighbor at*

SATURN
Diameter: 75,000 miles
Distance from sun:
887 million miles
Period of solar orbit:
29.5 years
Rotation period:
10.3 hours
Moons: 10

URANUS
Diameter: 29,580 miles
Distance from sun:
1.8 billion miles
Period of solar orbit:
84 years
Rotation period:
10.7 hours
Moons: 5

NEPTUNE
Diameter: 27,590 miles
Distance from sun:
2.8 billion miles
Period of solar orbit:
164.8 years
Rotation period:
15 hours
Moons: 2

PLUTO
Diameter: Approximately
3,700 miles
Distance from sun:
3.7 billion miles
Period of solar orbit:
248.4 years
Rotation period:
6.4 days

PAINTING BY RICHARD SCHLECHT

110,000 miles an hour; Pluto crawls at one-tenth that speed. A total of 32 moons orbit six of the planets. The asteroid belt (brown band in diagram above), holding more than 50,000 chunks of planetary material with diameters up to 470 miles, rings the sun between the orbits of Mars and Jupiter. Some scientists speculate that these planetoids once made up a larger planet, now fragmented. Others believe that the particles never had a chance to coalesce into a planet because Jupiter's heavy gravitational pull kept them scattered. Except for Pluto, inclined more than 17°, the planets orbit the sun in approximately the same plane (above). This oblique-angle view exaggerates the slightly elliptical paths of the planets.

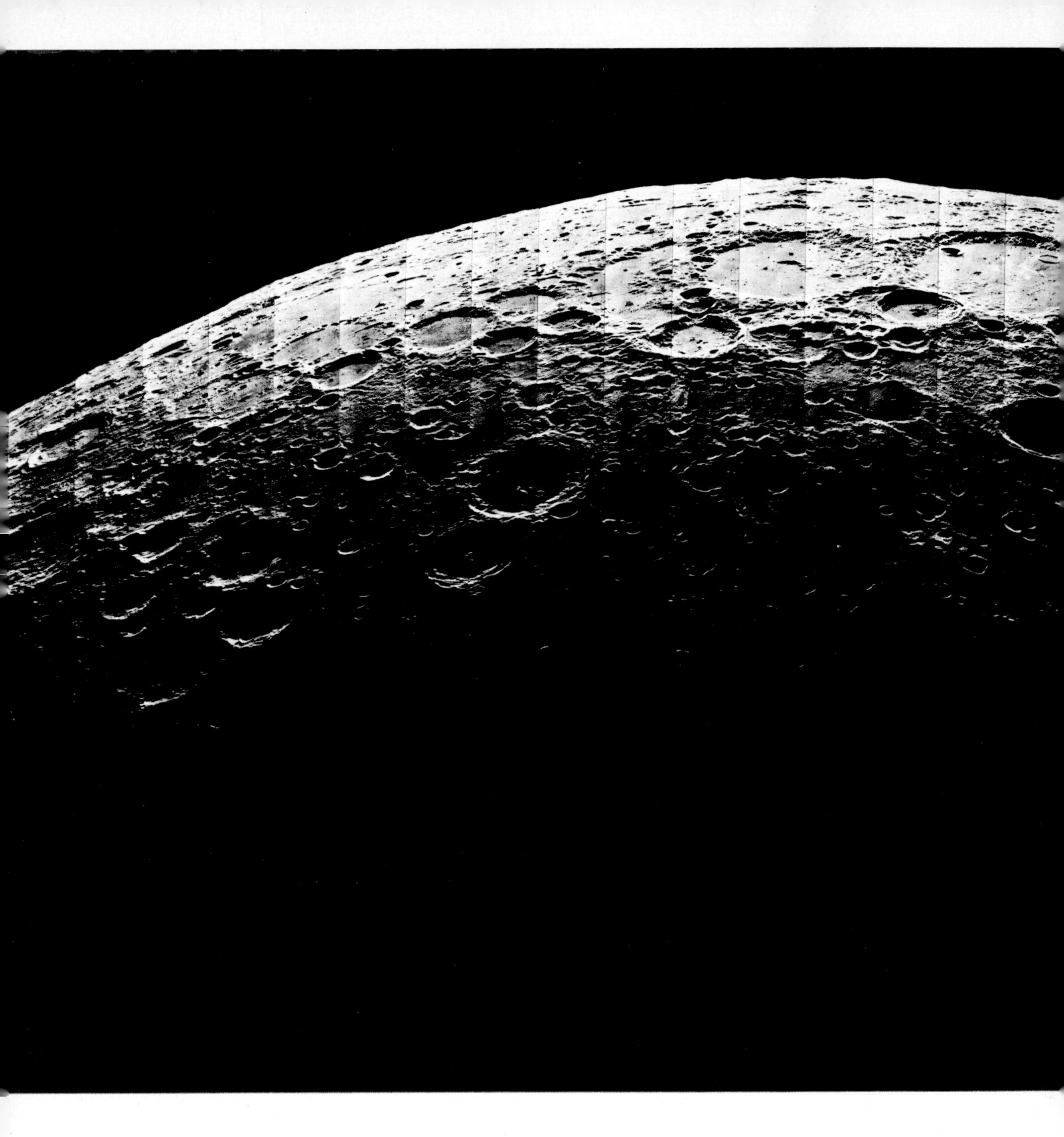

muscles had to force his arms to raise and keep his body in front of the craft as he tried to hook it by a tether to the Agena rocket. Finally, in desperation, he climbed astride the spacecraft and wedged his legs between it and the docking nose. "Ride 'em, cowboy!" Conrad called to him by radio. Minutes later, Conrad eased the nose of his craft into the docking collar of Agena.

The astronauts found and docked with the rocket during their first orbit, a record for making a rendezvous. Thrust into space at the precise second that would make the job easy, the men calculated their rocket-firing sequence with a compact computer and maneuvered right to the target.

Conrad and Gordon's performance was almost perfect. But if they had started the first burn of the thrusters even a second too early or too late, or let it go on too long or too little, the spacecraft would have missed by miles the exact point necessary for the success of the next burn. Corrective moves would have used up fuel intended for other jobs on the flight.

On the third day, the two astronauts had another chance to experience orbital physics—even to worry about it—and to stick their heads into the bottom edge of the Van Allen belt. After locking the nose of their craft into the Agena, they fired its rocket, propelling themselves about 675 miles higher. A jolt of acceleration and the gravity it created interrupted their weightless state. "Whoop-

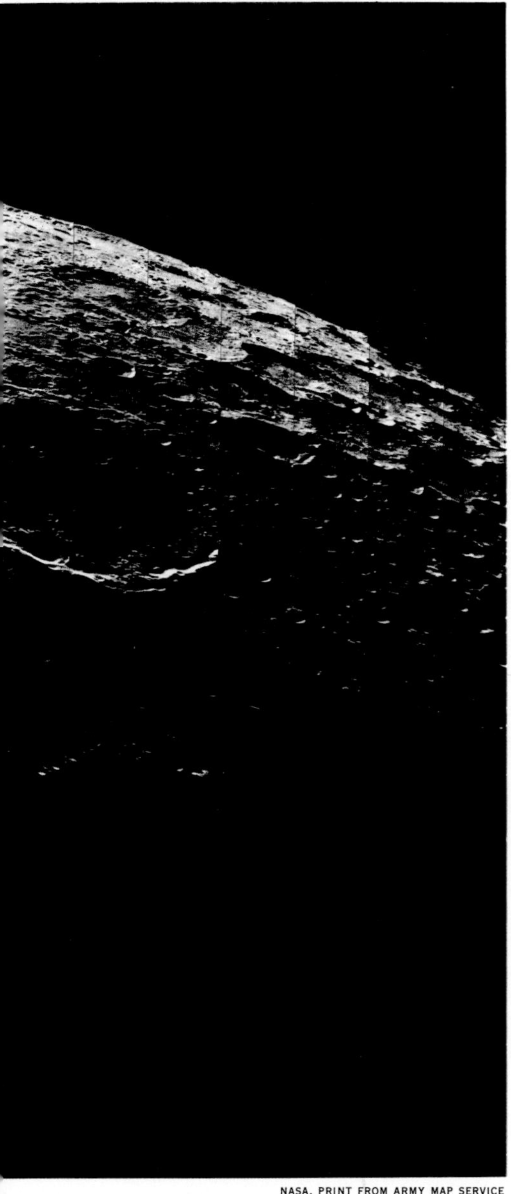

NASA, PRINT FROM ARMY MAP SERVICE

SUNLIGHT *reveals a stark moonscape perpetually hidden from earthbound observers in this photograph taken by Lunar Orbiter 5 on August 11, 1967. Lacking the protective shield of earth and its atmosphere, this surface presents an easier target for meteorites and solar particles than does the earth side. Astrogeologist Eugene M. Shoemaker, working in Flagstaff, Arizona, produces detailed maps of the moon's geologic and topographic structure from thousands of lunar photographs.*

FRED WARD, BLACK STAR

de-doo!" laughed Pete Conrad. "Look at it go!"

The new orbit attained an apogee, or high point, about 853 miles above earth. Its perigee, or low point, was 185 miles. The craft speeded up at the low end of the orbit to nearly 18,000 miles an hour, the greatest speed attained by man up to that time. At the high end, over the suburbs of Brisbane, Australia, it slowed to about 15,000 miles an hour. The longer sweep upward and the slower speed allowed the craft to "linger" a few minutes over Australia and the Pacific and Indian Oceans.

The astronauts made two of these lopsided flights before the Agena rocket shot them back to their former circular orbit.

At a high point on their first elliptical orbit, they had opened the hatch. Gordon stood up, his head and shoulders in space, his weightless tethered body floating a little above the floor. For two hours he took pictures of the earth—great reaches of oceans, whole mountain ranges, vast pink deserts, heavy green jungles—and the stars.

During two "night" periods of less than an hour each, the astronauts searched out the stars Antares and Canopus and the constellations Scorpio and Orion. The stars shone with a bright, steady light. The twinkle caused by earth's busy, shimmering atmosphere no longer blurred their outlines.

During a short lull in their picture-taking, both unexpectedly dropped off to sleep, Gordon with his head still out the hatch! Why not? There was

© PARIS-MATCH, FROM PICTORIAL PARADE

ADÉLIE PENGUINS *appear unperturbed during the launch of a French Dragon rocket in 1967 from the Dumont d'Urville base near the South Magnetic Pole. The 28-foot-long solid-fuel rocket, fired to an altitude of 200 miles as part of France's program of Antarctic exploration, recorded electron density and temperatures in the ionosphere.*

no noise to bother them; without air, space is completely silent except for radio voices piped inside the helmet. Conrad woke first from his catnap and called to his companion. "Huh? Oh . . ." said Gordon, and aimed the camera again.

In the regions where the astronauts were orbiting, they could expect to hear an occasional tiny ping when a speeding speck of matter about half the size of a grain of sand struck their craft. "Astronauts who have heard the impacts say they sound like bird shot pellets on a tin roof," says astrophysicist Curtis L. Hemenway, director of Dudley Observatory, Albany, New York. Sample squares of material that Dr. Hemenway arranged in boxes attached to long-orbiting satellites have recorded the dents left by micrometeoroids.

An international group of experts has agreed, Dr. Hemenway says, that the impact rate is about one micrometeoroid every hundred seconds on a surface of about one yard square. For a craft the size of Gemini 11, that would amount to about a million pieces of microscopic dust a month and perhaps a hundred barely visible specks. The chances that during a month a spacecraft would be struck, perhaps pierced, by a meteoroid as big

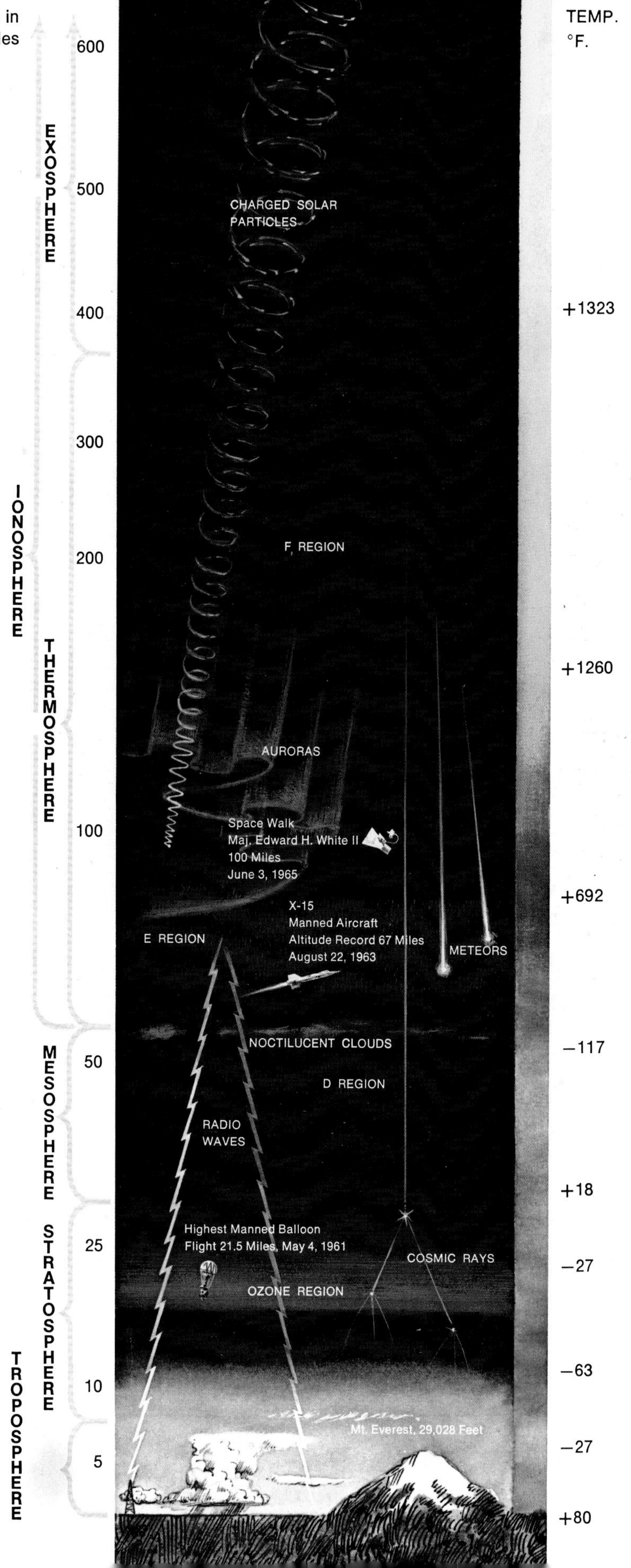

EARTH'S ATMOSPHERE, *a mixture of gases, water, and dust, consists of five major regions of varying pressure, temperature, and density. The troposphere, ranging seven miles high, contains 75 percent of our atmosphere; here occur clouds, storms, and fog. In the stratosphere, ozone, a form of oxygen, filters out much of the sun's ultraviolet radiation. Noctilucent clouds, possibly specks of dust coated with ice, drift high in the mesosphere. Several kinds of gases in the thermosphere absorb ultraviolet rays and solar particles that, when intense, cause auroras. The atmosphere merges with interplanetary space in the exosphere.*

The ionosphere, composed of charged atoms and electrons, and molecules, reflects radio waves beamed from earth. The F region reflects shortwave emissions; normal broadcasting frequencies reach the E region. The D region absorbs radio signals and after intense solar activity can cause severe communications blackouts.

Cosmic rays, high-energy particles from space, collide with nuclei of air molecules, producing some secondary rays that reach earth.

The temperature falls rapidly as altitude increases, until warming slightly about 30 miles up. In this region, ozone absorbs ultraviolet radiation, a reaction that produces heat. In the mesosphere, the ozone content thins out, and temperatures again decrease only to rise markedly as ultraviolet radiation excites, or ionizes, gas molecules.

PAINTING BY RICHARD SCHLECHT

Altitude in Statute Miles

TEMP. °F.

600

EXOSPHERE

500

CHARGED SOLAR PARTICLES

400

+1323

300

IONOSPHERE

200

F REGION

THERMOSPHERE

+1260

AURORAS

100

Space Walk
Maj. Edward H. White II
100 Miles
June 3, 1965

+692

X-15
Manned Aircraft
Altitude Record 67 Miles
August 22, 1963

E REGION

METEORS

NOCTILUCENT CLOUDS

−117

50

MESOSPHERE

D REGION

RADIO WAVES

+18

25

Highest Manned Balloon
Flight 21.5 Miles, May 4, 1961

COSMIC RAYS

−27

STRATOSPHERE

OZONE REGION

10

−63

TROPOSPHERE

Mt. Everest, 29,028 Feet

5

−27

+80

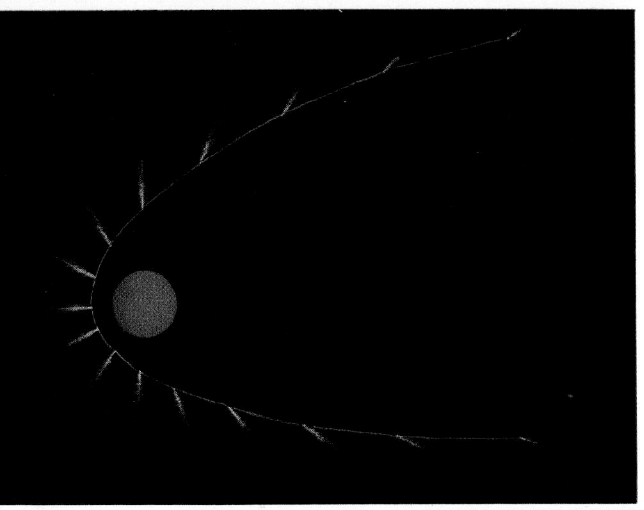

DRAWING BY ROBERT W. NICHOLSON

SCARRING THE DAWN SKY *near Washington, D. C., the comet Ikeya-Seki (opposite) stretches its 75-million-mile tail starward during its 1965 sweep past the sun. The comet closed to within 300,000 miles of the solar surface, at speeds up to a million miles an hour. Composed of gases, porous solids, and ice, comets heat and luminesce as they approach the sun. The tail forms from cometary material swept back by the charged particles of the solar wind. This force exerts a pressure strong enough to push the tail away from the sun's surface (above). Below, Saxons point fearfully toward Halley's comet, which appeared in 1066, a few months before the Norman invasion of England. This 11th-century Bayeux Tapestry interprets the visitation as an evil omen. Visible on earth at intervals of 76 years, the comet will return during 1986.*

MILTON A. FORD AND VICTOR R. BOSWELL, JR., N.G.S. STAFF (BELOW), BY SPECIAL PERMISSION OF THE CITY OF BAYEUX; VICTOR R. BOSWELL, JR. (OPPOSITE)

as a garden pea are only about one in a million.

When a meteoroid reaches earth's atmosphere, it may be traveling as much as a hundred thousand miles an hour. Superheated air in the shock wave it creates burns the meteoroid—now a shooting star—just as Conrad and Gordon's craft would burn if it had no heat shield when it re-entered earth's atmosphere. Bits of microscopic dust, on the other hand, like tiny parachutes, slow down and float gently to the ground. About a thousand tons of dust fall on us every day, judging from the evidence collected by spacecraft.

Some dust comes from our natural satellite. "Little parts of moon are falling on us all the time," says astrogeologist Eugene M. Shoemaker. This source, the moon itself, was the destination of Conrad, Gordon, and Bean in November 1969.

After a pinpoint landing, Conrad became the third man to step to the lunar surface. "That may have been a small step for Neil, but it's a long one for me," quipped the 5-foot-6½-inch astronaut.

After Alan Bean joined him on the surface, some of the work performed—including that of instruments they left there—helped determine several physical properties of the celestial body nearest earth: The moon does have an atmosphere, though very thin; it has more magnetism than anyone had thought possible; the crust contains some apparently harmless radioactive elements; and in the absence of the wearing action of water, something else is nevertheless "weathering" the rocks.

As the men worked, the nine layers of their tailor-made pressure suits provided adequate but temporary protection from the solar wind. Had earth instruments monitoring the sun reported a major and unexpected solar flare, however, they would have been ordered immediately inside their spacecraft and possibly back to earth.

On the Apollo 11 flight, Aldrin saw mysterious flashes of light, even with his eyes closed. It was later determined that this phenomenon, which some other astronauts also experienced, was caused by cosmic rays—mostly protons—striking the retina of the eye. On Apollo 12, similar particles left a record in the form of streaks found later in the astronauts' helmets, which were of somewhat different plastic than previous ones.

Although heavy shielding of manned spacecraft or helmets has not yet been used, the scientists who studied the Apollo 12 helmets suggested that

such shielding would be necessary on extended flights to avoid damage to an estimated 1.5 percent of the larger brain cells, and possible injury to the cerebral cortex and to the eyes.

Instruments the Apollo 11 and 12 astronauts left on the lunar surface included a solar wind spectrometer to measure the energy, velocity, and directions of charged particles of hydrogen, helium, and other elements streaming out from the sun; a magnetometer to measure the magnetic field and to indicate the presence of iron beneath the surface; an atmosphere detector to measure the minute pressure near the surface; an ionosphere detector that can identify traces of such gaseous material as argon 40 given off by elements in the lunar surface; and a seismometer, which has shown that our natural satellite has periodic "moonquakes," usually when it is under the strongest influence of earth's gravity.

During their return to earth, the Apollo 12 crew became the first humans to witness an eclipse of the sun by the earth. As all earth features except shape became lost in the darkness, the astronauts watched and photographed a tiny sliver of the sun as it revealed a never-before-recorded impression of our atmosphere. Alan Bean declared it "the most spectacular view of the whole flight."

Some instruments left behind have already increased man's knowledge of the solar system.

We now know, for example, that the differences between Mars and Venus—like the differences between the earth and its moon—are more striking than the similarities. Each celestial body we have explored so far has given up secrets that seem to say, "I am unique among all the others." This makes fitting components of our solar system into one "grand design" a more difficult and, at the same time, a more fascinating challenge.

Beyond Mars lies the asteroid belt, where great chunks of rock fly around the sun. At least 50,000 have been detected, from mile-wide pieces to spheres and irregular lumps up to 470 miles across. Occasionally, one falls to earth as a meteorite. Some authorities believe that asteroids are fragments of an exploded planet. Others think that Jupiter's gravitational force may have prevented a planet from forming. Or perhaps small planets developed, then collided and shattered.

The eccentric orbits of some asteroids, which periodically bring them relatively close to earth, may someday permit a visit by a spacecraft. Icarus, half a mile in diameter, has come as close as 4 million miles; Geographos (about the same size), named for the National Geographic Society, swings within 5.6 million miles; Toro, 1.5 miles across, comes within 9.3 million miles.

If astronauts ever attempt a flight to Jupiter, they will have to dodge the hurtling obstacles of the asteroid belt. Other space debris, perhaps left by comets, might also plague the spacemen.

Hundreds of known comets journey around the sun, usually in extremely elliptical paths coming close to the sun at one end, and swinging far out beyond the orbit of Pluto at the other. Their enormous cloudy heads measure as much as several hundred thousand miles in width. Scientists believe that in many comets, this cloud surrounds a ball of rock- and dust-filled frozen gases sometimes 50 miles in diameter. Near the sun, some of the gases in the "snowball" thaw and cause the cloud around it to spread out spectacularly.

A tail forms when the solar wind particles strike the cloud. The tail, as much as 100 million miles long, gradually disappears as the comet moves far away from the sun. Most astronomers think small meteoroids and micrometeoroids are debris

SOL GOLDBERG, CORNELL UNIVERSITY

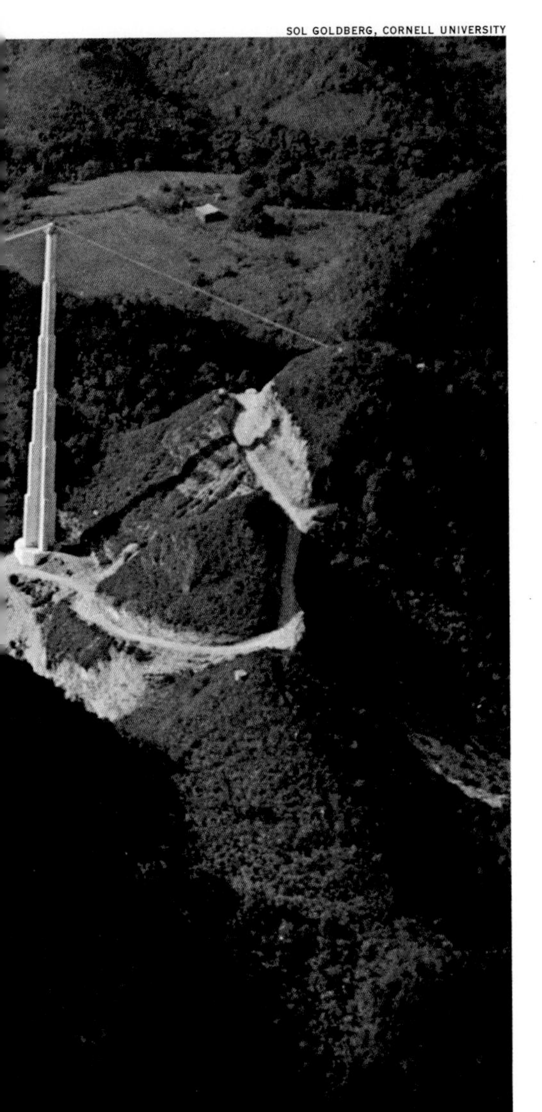

COMMONWEALTH OF PUERTO RICO

GIANT TELESCOPIC EAR, *the 1,000-foot-wide Arecibo Observatory of the National Astronomy and Ionosphere Center in Puerto Rico explores the universe by sending and receiving radio waves. Among its achievements: mapping the surfaces of Mercury, Venus, and Mars. Aluminum panels will replace the wire mesh of the telescope's dish, increasing its power and permitting it to chart these planets more accurately. Crossing the dish, a technician wears water skis to distribute his weight.*

J. R. EYERMAN

ANDROMEDA *(opposite), sister of our Milky Way and one of billions of galaxies in the universe, spins in space some 2.1 million light-years away. The hub glows yellow with ancient stellar giants; young blue stars shine among the spiral arms. Two smaller galaxies hover nearby. The 200-inch Hale telescope (above), largest of three scanning the heavens at California's Palomar Observatory, penetrates to the limits of observable space.*

scattered along the comet's path. When the earth crosses that path, some debris plunges into our atmosphere and the night sky blazes with a flurry of shooting stars.

Myriad questions remain about what is out there in the vast reaches of space, what is going on. Perhaps the most intriguing question to man is that of *who* is out there—if anybody.

Pete Conrad would probably take off tomorrow to look for the answers if there were a craft ready to go. He is the sort of man whose curiosity and sense of adventure would urge him on to get a close look at a planet, at a star, at another galaxy.

But the rockets available to him now wouldn't get him far. Our most powerful, the Saturn V, with 7.5 million pounds of thrust, could take 140 tons of payload into orbit 120 miles high. Several such payloads, assembled in orbit into a nuclear-powered rocket, could go to Venus or Mars—but not much farther—and return within two years.

Two new types of engine, now used in experimental flights in both the U. S. and Russia, provide exhaust thrust with beams of highly accelerated mercury ions. These engines require stores of fuel small enough to be practical for long journeys once underway. The thrust of each engine is small, but it can build up high speeds for craft in frictionless space.

In the ion engine, beams of positive ions accelerated by an electrical field exhaust from the rear of the engine as they mix with a stream of negative electrons. On a similar principle, the plasma engine uses electromagnetic forces to accelerate both ions and electrons into an exhaust beam.

Both rockets require great quantities of electricity, however, and the speeds they achieve would be far below the speed of light.

The distance to even the nearest star, around which hospitable planets may or may not be orbiting, is measured in light-years, the distance light travels in a year—about six trillion miles. Obviously, if man wants to go into stellar space, he must travel as close to the speed of light as he can or else he will never be able to make the trip out and back in a lifetime. According to Einstein, the velocity of light cannot be exceeded. Man could never go that fast, he said.

But some dreamers think man can travel almost that fast, and they suggest the photon rocket as a likely means. Scientists know that light exerts pressure when it strikes a surface. Even now some suggest that astronauts on journeys to Venus or Mars could unfurl enormous light-reflecting sails, catch the pressure of light from the sun, and eventually travel at high velocities without using any fuel! The force produced by powerful beams of photons shooting from a rocket might succeed in accelerating the craft to velocities approaching the speed of light, 186,000 miles a second.

But would Pete like going 6,000 trillion miles to a star if, according to the time scale Einstein said would prevail during his trip, he came back only three decades older to find that the earth had aged more than 2,000 years! As Dr. Von Braun says, "He might wind up in a zoo."

No, thanks, Pete might understandably reply. Then again, he might repeat the remark he made soon after his Gemini 11 flight: "Give me a good night's sleep, and I'm ready to go back again."

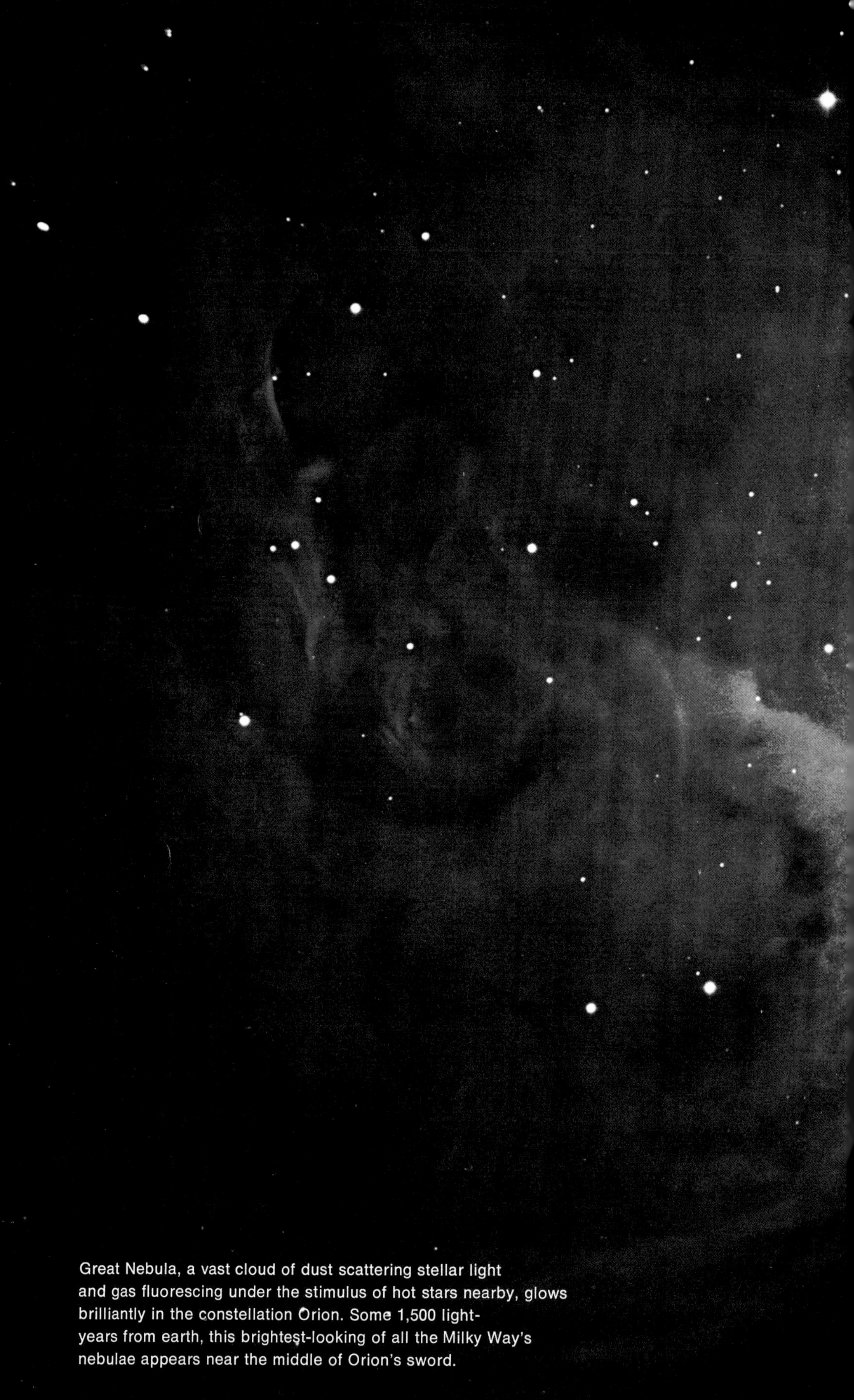

Great Nebula, a vast cloud of dust scattering stellar light and gas fluorescing under the stimulus of hot stars nearby, glows brilliantly in the constellation Orion. Some 1,500 light-years from earth, this brightest-looking of all the Milky Way's nebulae appears near the middle of Orion's sword.

4/ THE ALIEN ENVIRONMENT: MAN LEARNS TO ADJUST

On an unusually warm Florida morning, some two hours before Apollo 11 roared aloft on its historic mission to the moon, I noticed a slender man dashing about the crowded press site. As he snapped pictures with an inexpensive camera, he looked at first very much like an excited tourist. A few moments later when our paths crossed and we recognized each other, the busy photographer turned out to be a rare "tourist" indeed—a man who had envisioned more journeys into space, probably, than any other person there. He was Arthur C. Clarke, distinguished British scientist and science fiction writer, father of the communications satellite, and co-author (with Stanley Kubrick) of the movie *2001: A Space Odyssey.*

As we discussed the impending effort to land the first men on the moon, I reminded Clarke of an analogy he had once drawn regarding man's drive to conquer the cosmos. The departure from earth, Clarke had suggested in his book *Voices from the Sky,* could be compared to life's emergence from the sea—what he called "the perfect environment for life"—to invade the new and formidable province of land and air.

"In the sea," Clarke wrote, "an all-pervading fluid medium carries oxygen and food to every organism. . . . The same medium neutralizes gravity, ensures against temperature extremes, and prevents damage by too-intense solar radiation—which must have been lethal at the Earth's surface before the ozone layer was formed.

". . . it seems incredible that life ever left the sea, for in some ways the dry land is almost as dangerous as space. . . . We seldom stop to think that we are still creatures of the sea, able to leave it only because, from birth to death, we wear the water-filled space suits of our skins."

To neither of us, on that morning of July 16, 1969, did it seem unnatural that man now had the capability of journeying some quarter of a million miles from his planet to make the first footprints on another body in space. Yet less than a dozen

WILD RIDE *in a bizarre training chair—used by some Mercury astronauts—tests a pilot's ability to stabilize a dizzily careening spacecraft. A control stick counters roll, pitch, and yaw. Engineers at the Lewis Research Center in Cleveland, Ohio, attached lights to the frame to trace its tumbling.*

NATIONAL GEOGRAPHIC PHOTOGRAPHER DEAN CONGER

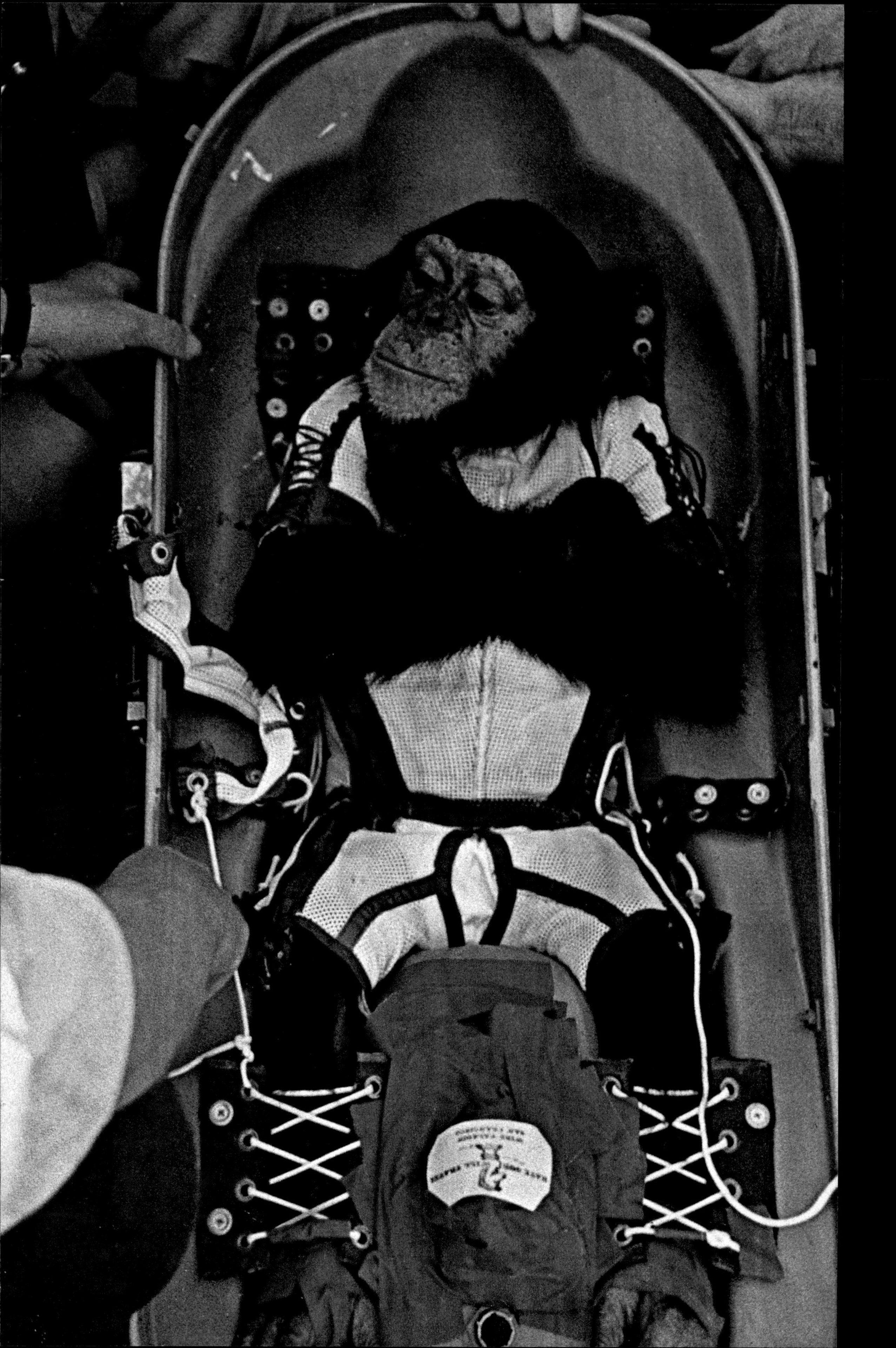

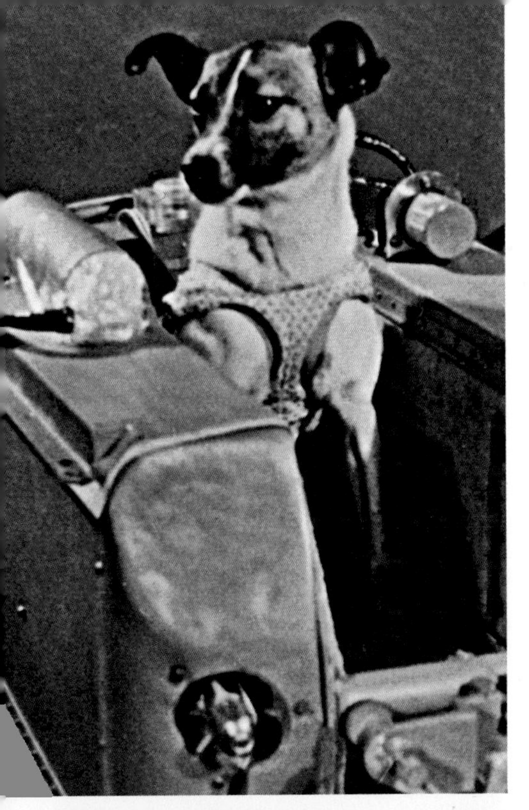

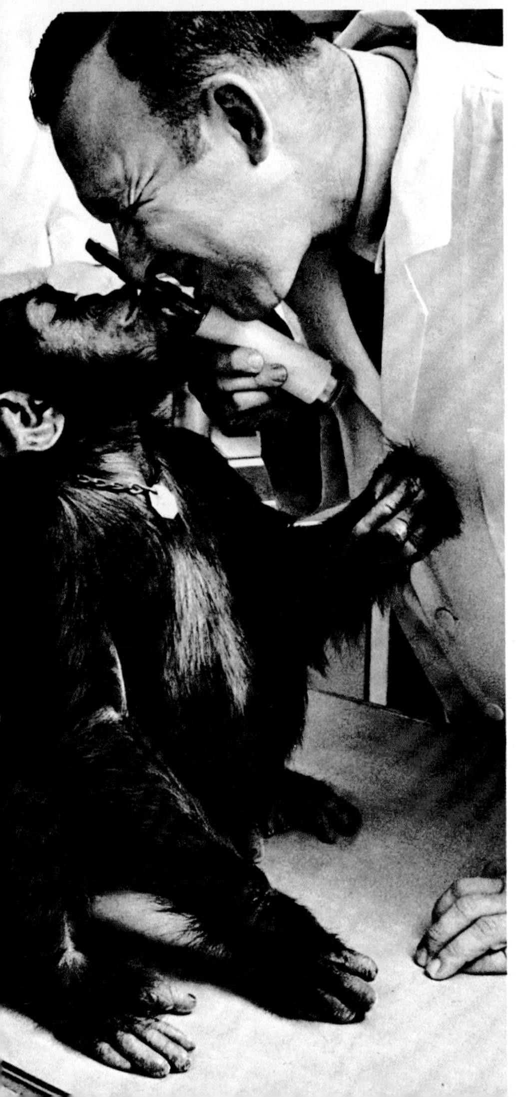

years earlier, many experts had voiced doubts that man could ever adapt to a weightless environment. Numerous biologists and physicians had expressed fears that muscles might atrophy if relieved of the stress of gravity for long periods. Others had predicted that either high or low blood pressure would result when the heart was suddenly freed of its normal task of pumping against gravity. The inner ear caused concern—the delicate mechanism that enables us to detect motions and acceleration, and to keep our balance. Would the weightless state so confuse it, doctors wondered, as to deprive an astronaut of his sense of orientation? If a man tried to drink water in space, some asked, wouldn't the weightless fluid stick in his gullet and choke him?

There were also widespread fears that micrometeoroids might puncture and disable spacecraft, even though the President of the Soviet Academy of Sciences, Mstislav Keldysh, insisted the danger was no greater than "the probability of a brick falling on your head as you walk along the street."

Part of this initial scientific caution was undoubtedly based on the healthy respect man had acquired in recent centuries for space as a hostile region. Recognition of its hazards dates only from about the mid-17th century. Until then most people assumed—if they thought about it at all—that the air they breathed extended all the way to the moon, and probably far beyond.

Lucian of Samosata, a fantasy writer who lived 1,800 years ago, sent his passengers to the moon on a ship powered by winds. It was Johannes Kepler's fictional account of a moon trip about 1630 that first suggested a lack of air and gravity in space—anticipating by 60 years Isaac Newton's formulation of the law of gravity.

High-altitude balloon experiments, beginning about 1800, provided evidence of the lethal cold

FRECKLED CHIMP HAM, *in 1961 a forerunner of man in space, awaits release from his couch after a 16-minute, 5,800-mph flight in a Mercury capsule. Fired aloft by a Redstone rocket, Ham soared 156 miles high and splashed down in the Atlantic 414 miles from his launching pad at Cape Canaveral. Nuzzling a doctor, astrochimp candidate Billy sits for an eye examination. The Russian dog Laika, first animal to orbit the earth, died aboard Sputnik 2 in 1957 after remote sensors monitored her pulse, respiration, and other bodily responses.*

HENRY BURROUGHS, ASSOCIATED PRESS, NASA (FAR LEFT); NOVOSTI (TOP); RALPH CRANE, LIFE © TIME INC.

and asphyxiating vacuum awaiting the space traveler. In 1804 Frenchman L. J. Gay-Lussac recorded sub-zero temperatures at 23,000 feet, and in 1862 James Glaisher of England blacked out at 29,000 feet. A few years later, two French balloonists died of oxygen starvation at 27,950 feet, the first casualties of high-altitude flight.

In the age of the airplane, man overcame the hazards. He wore warm clothing; he donned an oxygen mask. Eventually he took along his own comfortable atmosphere by pressurizing his aircraft. But since his new wings could take him only a few miles into the atmosphere, he could learn little of what lay beyond. For this he had to await the development of rockets capable of lifting him 30 miles or more.

At least one experimenter was ready to rocket a human skyward much earlier, however. In the 1830's Claude Ruggieri, former fireworks master for King Louis XVI of France, proposed to send up a small boy in a gunpowder rocket he devised. Although Ruggieri had succeeded in launching mice and rats, the Paris police forbade his bold experiment with a human being.

The launching of liquid-fueled rockets for biomedical experiments in space began after the United States gained possession of the German V-2's at the end of the second world war. The program eventually used chimpanzees, but university and military scientists first sent up such varied specimens as seeds, onions, yeast, sea-urchin eggs, fruit flies, mice, and monkeys.

Similar experimentation in the Soviet Union began in the 1940's and 1950's with white mice, frogs, even goldfish, but the Russians came to rely primarily on dogs. In Moscow in 1966 I asked a leading expert in bioastronautics, the late Academician Vasili V. Parin, why this was so.

"The Russian dog," the gray-haired biologist replied, "has always been a great friend of science. You know, of course, of the work of our famous psychologist, Ivan Pavlov, who studied the reflexes of dogs. But did you know that near Leningrad there is a monument in honor of Pavlov's dogs? We have amassed much data on our four-footed friends. Their circulation and respiration are close to that of man's. And they are very patient and durable under long experiments."

When the Russians decided to send Laika into orbit in 1957, they made no provision for bringing her back. Their main goals were to discover

N.G.S. PHOTOGRAPHER OTIS IMBODEN (ABOVE); NASA

CHOCO INDIAN *of Panama watches as Astronaut R. Walter Cunningham eats spit-roasted iguana in 1964 during the jungle phase of a survival course. Four years later, Cunningham orbited the earth*

163 times aboard Apollo 7. In 1967, astronauts improvised protective clothing and a makeshift shelter from nylon parachutes on a four-day desert exercise in Washington State. Although trained to feed, clothe, and protect themselves in the event of an off-target landing in a remote area, all of the 30 United States flight crews splashed down safely at sea within reach of recovery vessels.

how live creatures would react to extended weightlessness, and to learn how to improve life-support systems. They were confident that if they could keep the dog alive for a week and record her responses, they would obtain the data they needed. The mission went as planned, and before Laika died, she demonstrated that living organisms could adapt to space flight.

In the meantime, the United States was working on the same problem. On May 28, 1959, two monkeys named Able and Baker rocketed aloft from Cape Canaveral, reached an altitude of 300 miles and a speed of 10,000 miles per hour, and traveled 1,071 miles downrange. After their safe return, scientists proceeded with plans for the launching and recovery of chimpanzees.

In 1961 the flight of a chimp named Enos proved him a remarkably stable character. At intervals during his two-orbit flight, colored lights flashed. If Enos then pressed a lever, as he was trained to do, he could obtain banana pellets and water—and avoid mild electric shocks. The action signaled ground stations that he was performing his duty while in flight.

But trouble developed in the circuitry, and even though America's first orbiter was prompt in pressing the lever he often got a shock anyway. In-flight photographs reviewed later showed Enos baring his teeth in exasperation; nevertheless, he continued to press the lever purposefully every time the light flashed.

I saw Enos at a press conference just after his recovery. The astrochimp struck me as poised and alert. Surprisingly, he still relished banana pellets.

Now, at last, man had reached the point of launching himself into space. Ages earlier, life had only one way to survive when it left the sea: adapt, through evolution, to such factors as gravity, air, and radiation. Today, man can insulate himself from the alien environment. Thus when the U.S.S.R. and the United States built their earliest spacecraft—Vostok and Mercury—their scientists and engineers were challenged to develop capsules that would withstand extreme cold and heat and the brutal stresses of launch, re-entry, and landing. And they had to provide a livable atmosphere within the craft and within the spacemen's pressure suits.

When I inspected the spherical Vostok in Russia, I saw a metal shell that appeared to be nearly two inches thick. The spacecraft seemed as solid as a locomotive. Layers of an asbestoslike material provided insulation.

The Mercury craft, by contrast, used a thin outer shell of cobalt-nickel shingles, a 1½-inch core of metallic insulation, and an inner shell of corrugated titanium. For their spacecraft atmosphere the Russians chose a basic nitrogen-oxygen mixture much like the air we breathe on earth. The United States used pure oxygen under 5.5 pounds pressure per square inch. Air-conditioning systems controlled the temperatures.

Both Vostok and Mercury worked amazingly well—and their pilots reacted to space flight with contagious enthusiasm. The first sign of an adverse reaction to weightlessness came on the second Soviet flight, that of Gherman S. Titov. Recalling his experience on entering orbit the morning of August 6, 1961, Titov wrote: "For the life of me I could not determine where I was . . . I was completely confused, unable to define where was earth or the stars. . . . Something had gone suddenly and drastically wrong with . . . my sense of balance."

Although Titov's disorientation lasted only seconds, dizziness and nausea recurred several times during his 17-orbit mission. His difficulties caused such concern that some experts in bioastronautics questioned whether man could function for prolonged periods in space.

Such fears moderated, however, when his reactions were not immediately repeated by other astronauts and cosmonauts. The first woman in space, Valentina V. Tereshkova, experienced nothing worse than a bruised nose, suffered as her spacecraft landed by parachute in 1963.

Then in October 1964, the Soviets orbited pilot Vladimir M. Komarov, physiologist Boris B. Yegorov, and physicist Konstantin P. Feoktistov in Voskhod 1, the first three-man spacecraft—and both Yegorov and Feoktistov briefly experienced Titov's kind of disorientation.

Gemini crewmen showed certain body changes: Bone density lessened slightly, minute quantities of calcium and protein nitrogen were lost, blood red-cell counts dropped 5 to 20 percent.

Blood pressure, however, remained within normal ranges, no muscular atrophy or serious loss of coordination occurred, and even the 14-day flight of Gemini 7 caused no motion sickness or disorientation. The bodies of the astronauts adapted to the space-flight stresses—confinement, restraint of pressure suits, 100 percent oxygen atmosphere,

ENCOUNTERING ZERO GRAVITY *for just under half a minute, astronauts float inside the fuselage of a jet tanker as it arcs over the roller-coaster crest of a parabolic curve. Charles A. Bassett drifts above Edwin Aldrin and Theodore C. Freeman; their instructor braces against the cabin wall.*

acceleration forces, weightlessness, vibration, dehydration, altered work-sleep cycles, tension, and a heavy workload. Drinking and eating in space proved to be no problem.

The functioning of the pressure-suited human body outside a spacecraft was tested during early EVA—extravehicular activity. The astronauts were initially plagued by perspiration that fogged helmet visors, and by excessive fatigue, but with modified pressure suits and improved techniques they demonstrated that man can work effectively outside the protection of his spaceship.

The United States decided that it was not necessary to precede manned moon flights with circumlunar visits by chimpanzees or other animals. But the Russians were far from satisfied that life forms in the vicinity of the moon could safely tolerate radiation caused by unexpected solar flares. (U. S. scientists believe these can be detected far enough in advance to give explorers on the moon sufficient time to return to the additional protection of their spaceship.)

Consequently Soviet scientists in 1968 sent around the moon two spaceships—Zond 5 and Zond 6—packed with dosimeters, which measure radiation. The two craft also contained turtles, wine flies, mealworms, a spiderwort plant with buds, and seeds of wheat, barley, and pine. It was only after the safe return of the life-forms in Zond 6 that the Russians concluded that radiation levels on such a flight were safe for man. At the same time they announced that Soviet scientists knew how "to create defenses against solar turbulences." Later I learned that these defenses included an antiradiation drug.

The U. S. was to demonstrate that man could survive and function on the moon without such a drug. Through the miracle of moon-to-earth television, millions actually saw it happen.

"There seems to be no difficulty in moving around," Neil Armstrong reported within a few minutes after taking that first historic step. Then the surprisingly quick physical adaptation was confirmed as Buzz Aldrin ran, drifted, hopped, and floated through a weirdly gaited choreography on the unfamiliar lunar stage.

During Apollo 11's elaborately controlled 21-day quarantine period, nothing harmful was found either in the astronauts' bodies or in the dust and rock samples they brought back.

Apollo 12 proved that Pete Conrad and Alan Bean could cheerfully conduct two lunar traverses totaling about 1¼ miles.

Man's physical limits in the new lunar environment were first indicated during the flight of Apollo 14, when Alan Shepard and Edgar D. Mitchell tried to move equipment up the steep slope of Cone crater. Shepard's heart rate reached 150 beats per minute, and Mitchell's 128, and the two prospectors had to turn back short of the rim.

Meanwhile, the Russians were discovering in their Soyuz series of earth-orbital flights some new justifications for their early caution. In late 1970, during the 18-day Soyuz 9 mission, both veteran Cosmonaut Andriyan G. Nikolayev and rookie Vitali I. Sevastyanov encountered puzzling difficulties—some of them new. After 24 hours in space, both reported impairment of vision and color perception, poor coordination of eye movements, and "a rush of blood to the head."

When I met the two cosmonauts in 1970, they reported that during the flight they experienced a mysterious loss of thirst and were unable to increase their heart rates above 100 beats per minute even after extended and vigorous exercise. They also revealed that for about the first eight days back on earth they had the unpleasant sensation of being at more than twice their normal earth weight.

So concerned were Soviet scientists over these reactions, and the possibility that the extended flight could make the cosmonauts more susceptible to infection, that the two men were quarantined for ten days after they landed. During that period, Nikolayev revealed, "there was some difficulty in maintaining vertical posture and in changing the gait. . . . The arms made involuntary movements to maintain equilibrium. . . ."

The next Soviet attempt to investigate these and other effects of prolonged weightlessness occurred on the record-breaking but ill-fated flight of Soyuz 11 in June 1971. For 24 days Lt. Col. Georgi T. Dobrovolsky, Vladislav N. Volkov, and Victor I. Patsayev orbited the earth in the 25-ton Salyut-Soyuz space station. Although they apparently suffered the expected physical and mental fatigue, the symptoms as far as they are known were much less severe and mysterious than those reported by Nikolayev and Sevastyanov.

On the 24th day Soyuz 11 was ordered back to earth. Even as the magnitude of the engineering and human achievement dawned on both East

BLAIR PITTMAN (BELOW); NASA

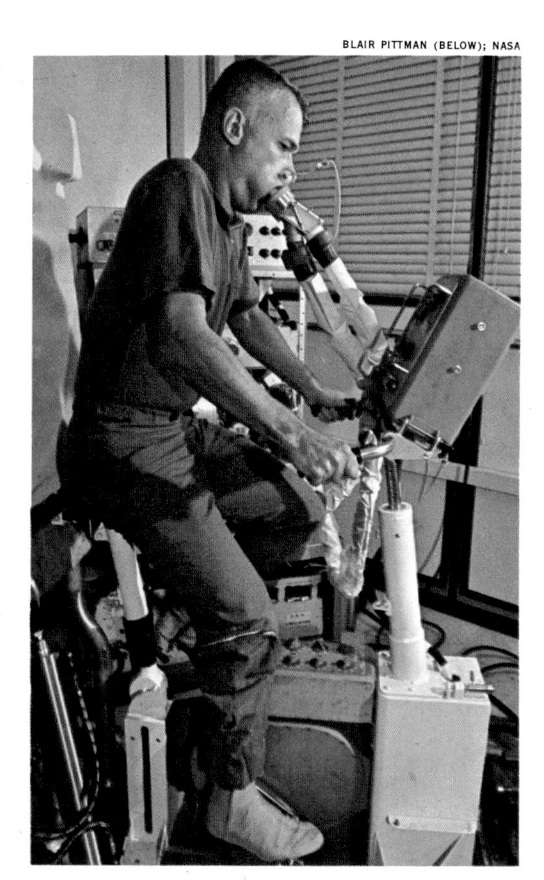

BACK FROM ORBIT *after a record-breaking 14 days aboard Gemini 7, James Lovell rubs a growth of beard, and Frank Borman waves to photographers waiting aboard the recovery carrier U.S.S.* Wasp. *In a space rendezvous on December 15, 1965, three days before the pair splashed down in the Atlantic, another twin-seater, Gemini 6, maneuvered to within a foot of them. Bruce McCandless II pedals an ergometer—an exerciser that measures the stresses of working in space—and breathes into a device that analyzes his metabolic responses. Sent aloft on May 14, 1973, Skylab served as America's first manned space station.*

RESCUE
RESCUE

and West, the world was stunned to learn that after a gentle landing the three cosmonauts were found dead in their seats. The cause of the tragedy, rapid loss of the spacecraft's atmosphere, was not related to effects of weightlessness.

On the next manned flight, that of Apollo 15, David R. Scott and James B. Irwin proved that men could perform strenuously in the one-sixth gravity of the moon for 18½ hours over a three-day period, provided they got enough rest and sleep.

It was apparent that after about eight hours of walking on the moon, both Scott and Irwin not only required less oxygen but also had gradually shifted their technique of locomotion to a new, energy-conserving, gliding action.

Alfred M. Worden, command and service module pilot, remained in the spacecraft 12¼ days without difficulty.

Back on earth, however, all three men took longer to return to normal than any previous astronauts. Irwin reported experiencing dizziness and several periods of irregular heart rhythm, both during and after the flight. Five days later, he still had the sensation when lying down that his head was much lower than the rest of his body. Scott for a time had double heartbeats.

Postflight examinations of Scott and Irwin showed an approximate 15 percent loss of potassium, a chemical essential to cardiac rhythm. Three days before the Apollo 16 lift-off, crewmen John W. Young, Charles M. Duke, Jr., and Thomas K. Mattingly II began eating a high-potassium diet, including potassium-spiked orange juice. They were to continue this diet throughout the mission and for three days thereafter. Although Young and Duke both had high heart rates while on their lunar excursions, neither experienced erratic heartbeats during or after their mission.

On the final lunar flight, that of Apollo 17 in December 1972, Astronauts Eugene A. Cernan and Dr. Harrison H. (Jack) Schmitt spent a record-breaking three days and three hours on the moon's surface. By then improvements in suit design, diet, inflight exercise regimens, and methods and schedules for work on the moon, had virtually eliminated medical restraints. Schmitt, Cernan,

HEADING INTO QUARANTINE, *the first men to fly a moon-landing mission walk toward a mobile unit aboard the carrier U.S.S.* Hornet. *When the ship reached Hawaii, a jet carried the Mobile Quarantine Facility to Houston; there the astronauts entered special quarters at the Manned Spacecraft Center for 18 more days of isolation. NASA abandoned this procedure after the third moon landing.*

and command module pilot Ronald E. Evans, who orbited overhead, reported no difficulties more severe than Evans's stomach cramps just prior to his final EVA.

The three long-duration Skylab missions, beginning in May 1973, opened an entirely new chapter in man's adaptation to weightless flight. Prior to Skylab, there had been speculation that the adverse reactions experienced by Soviet cosmonauts may have been caused by their relatively roomy spacecraft, which gave them increased mobility.

Now, with Skylab, we would have a chance to test man in almost hotel-like accommodations. The Apollo command module contained 219 cubic feet. Over-all, the 118-foot-long, two-story Skylab contained 11,300 cubic feet—about the same as a small house. The main wardroom alone measured 700 cubic feet. Special body and foot restraints were attached to the floors, cylindrical walls, and ceiling of the spacecraft. Special shoes were designed to anchor to the surfaces. The astronauts also were to have such luxuries as a book-and-music library (including the theme from the movie *2001—A Space Odyssey*), a shower, a washbasin, an air suction toilet, fireproof playing cards with magnets to hold them in place, a Velcro dart board, a zip-up private sleeping compartment, and a table where they could eat hot food from trays, cafeteria style. Everything, in fact, except a laundry room.

"The zero-gravity laundry development is something we haven't mastered yet," said Larry Bell, an engineer at the Johnson Space Center. Dirty clothes are thrown into the trash airlock.

Among Skylab's major objectives were the study of earth resources and of the sun; but from the beginning, another major Skylab objective was man himself. The first Skylab crew appropriately included the first U. S. physician in space, Dr. Joseph P. Kerwin.

The 98-ton Skylab space station was put into orbit empty. During launch it lost a white-painted micrometeoroid-thermal shield and one wing of solar panels. Aloft, it failed to deploy the other solar wing. After the situation had been studied on the ground for more than a week, mission commander Charles (Pete) Conrad, Jr., Dr. Kerwin, and Comdr. Paul J. Weitz rode the heaviest U. S. command module ever launched (13,295 pounds) to a rendezvous in orbit on May 25.

The engineering problems they encountered were serious. Unless Skylab's interior could be cooled, neither man nor food could survive. At one point the temperature had reached 125° F., enough, they later discovered, to burst most of the toothpaste tubes and to ruin all the tomato ketchup on board.

Initially, all tasks were subordinated to the engineering problems of freeing the jammed solar panel and making the crippled Skylab habitable. In one of the most brilliant episodes of space activity, the Skylab astronauts installed an orange parasol to shield the spacecraft from the scorching heat of the sun and freed the solar wing. These improvisations restored the interior temperature and

battery power to near normal. Now the real work could begin. What originally had been billed as a repair mission now became a productive work mission.

During the 28 days of the first manned Skylab flight, Dr. Kerwin, using the most sophisticated medical equipment ever taken into space, took astronaut temperatures, measured cardiovascular responses, and monitored sleep cycles and exercise on an onboard bicycle.

"None of us have had any motion sickness," reported Conrad. "It's obvious to me that man can work up here." His only complaint concerned the noise of the plumbing.

After splashdown on June 22, the astronauts were somewhat unsteady, but none was severely incapacitated by the sudden return to gravity.

The next crewmen to occupy Skylab, Navy Capt. Alan L. Bean, Dr. Owen K. Garriott, and Marine Lt. Col. Jack R. Lousma, were scheduled for 59 days in space. On the astronauts' advice they ferried up arm and upper torso exercisers to improve circulation and to help maintain muscle tone. Lift-off was July 28.

After they deployed a new sunshade, an oxidizer leak from one of their thruster valves confronted NASA with the possibility of having to launch an emergency rescue mission. An Apollo was modified to seat five and pressed into standby service. The uneasy Skylab astronauts found some diversion in watching the adaptation to zero-g of a now-famous spider named Arabella, an experiment that had been suggested by a high school student in Lexington, Massachusetts. Arabella messed up her first web but did much better on her second try. "She's a very fast learner indeed," observed Garriott.

Skylab's leak problems eventually moderated, and the rescue mission was never needed. Once the crew settled down to regular eating, exercise,

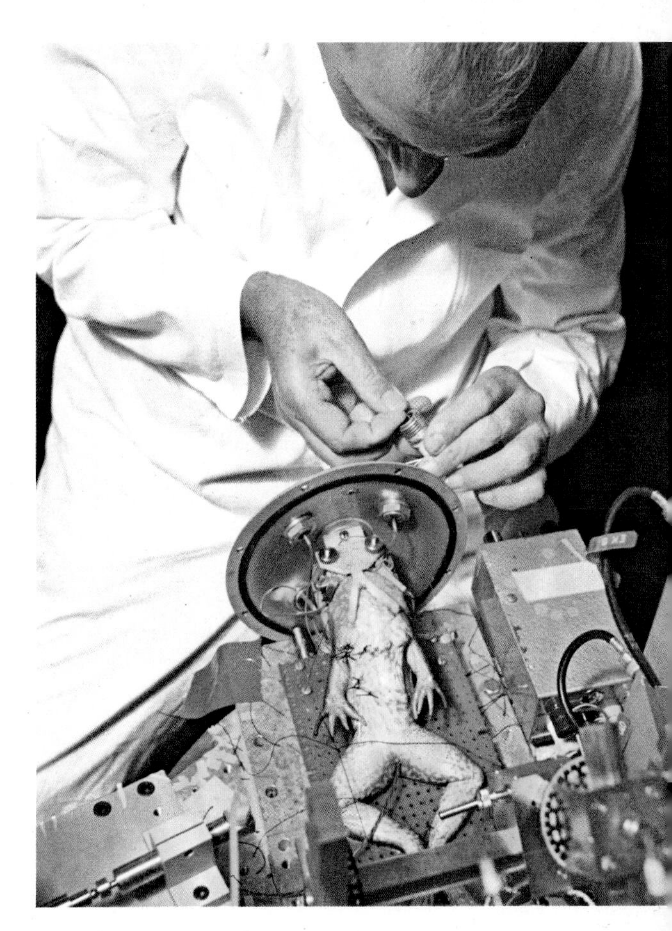

PREPARING A BULLFROG *for a test of responses to space flight, a neurophysiologist implants electrodes in its ear canal in a laboratory at Wallops Island, Virginia. Ground instruments in 1970 recorded the frog's reaction to acceleration, vibration, and zero-g. Below, a hamster breathes in fluorocarbon fluid at Emory University in Atlanta, Georgia. A tube attached to its heart monitors circulation, recording any changes caused by weightlessness. Dissolved oxygen and a lowered body temperature permit the hamster to survive in the liquid as long as 24 hours without harm.*

SOYBEAN PLANT *undergoes a sustained study of its reaction to a sprinkling of moon dust at Houston's Lunar Receiving Laboratory. Grown for 60 days in wood pulp containing nutrients, the plant showed no significant change when scientists added lunar material. But in other tests, minerals in the moon dust enhanced the growth of some of the plants. The chamber-within-a-chamber and the use of special rubber gloves ensured a sterile environment.*

DANA SHELTON (LEFT); NASA (UPPER); JOSEPH T. JACKSON

and sleep schedules, the early stomach difficulties and motion sickness abated.

On September 25 the Skylab astronauts returned to earth, 59 days and 11 hours after lift-off. They reported feeling tired during the early period of readjustment to gravity. Tests revealed they had mild "space anemia": a loss of muscle tone and about a 12 percent loss in red blood-cell production. But, said Jack Lousma, "we all felt as though we could have just stayed there indefinitely." The three astronauts stressed the importance of rest, exercise, and regular meals.

"It became obvious after three or four days," Bean said, "that there was enough work up there . . . to work all the time and not ever eat. . . . that kind of upset us a little bit. . . . So we're certainly going to recommend that the next crew get their food schedule locked in and treat that as one of the more immovable programs of the day."

The problem of readjusting to gravity is sometimes best illustrated by little things. Back in his home, Lousma, for example, once let go of a bottle of after-shave lotion in midair, fully expecting it to stay there, as it would have aboard Skylab.

The final flight to Skylab had been originally scheduled for 60 days. But with the scientific excitement generated by the discovery of the approaching comet Kohoutek—combined with reasonably good health reports from previous crews—the last flight was rescheduled for a possible 85 days. Lift-off was November 16, 1973.

Riding aloft with Marine Lt. Col. Gerald P. Carr, Dr. Edward G. Gibson and Air Force Col. William R. Pogue, were 25 pounds of extra socks and underwear and 392 special high-energy bars.

When the all-rookie crew floated into Skylab, they received a mild shock: The previous crew had left their space suits stuffed with clothing and propped up as dummies in the wardroom and bathroom.

Mission Control, which had learned by now the importance of a break in routine on long-duration space flights, was in on the joke and asked cryptically as the crew sat down to eat, "Were you able to find enough food for six people?"

Of special benefit to the third Skylab crew members was a treadmill that enabled them to exercise the lower body muscles and to restore blood circulation to the legs.

After traveling 29.9 million nautical miles in 84 days, one hour, and 16 minutes, the Apollo ferry craft landed in the Pacific near San Diego on February 8, 1974. The crew members may have been rookies, but they passed, in turn, veteran astronauts Jim Lovell, Pete Conrad, and Alan Bean, each of whom had held the previous world record for time in space. Two weeks later in Houston, the crew reported feeling well, better, as it turned out, than had previous Skylab crews.

The most important news from their personal reactions was that after about a month in weightlessness they seemed to reach a leveling-off point in terms of fitness for work, mental attitude, work proficiency, weight loss, and degree of irritability.

According to Jerry Carr, readapting to earth was similar in some ways to adapting to space, "The first two or three weeks in zero-g . . . you're always thinking about getting from here to there. . . . after you've been there for a while . . . you begin to forget about that. . . . I do remember very distinctly on the ship that I was right back in the same boat in one g . . . how am I going to get from here to there without ricocheting off the walls. . . ."

Medical tests showed that, as the space between spinal vertebrae expanded, the men had a temporary height gain of up to two inches, and that body fluids were redistributed toward the upper part of the body. Most important, all nine Skylab astronauts, in time, adapted to weightless flight, then readapted to the gravity they were born in.

Finally, they showed no ill effect from radiation experienced as a result of solar flares. The reason they did not, according to James E. Milligan, a principal investigator, was that "we're inside of a magnetic bottle. . . . The earth's magnetic field tends to shield the astronauts in the orbit they're in."

"From what we've seen," said Dr. Story F. Musgrave, an astronaut-physician, "I don't think there is any limit on how long man can stay in space."

JET-POWERED *Astronaut Maneuvering Unit propels mission commander Gerald P. Carr in a 1973 test conducted at zero gravity inside the Skylab space station during the third manned flight. Carr experiments with both back-mounted and hand-held propulsion systems designed to give astronauts mobility during EVA's—extravehicular activities; the back-mounted unit proved easier to pilot. The Skylab missions showed that man can perform useful work under conditions of sustained weightlessness without suffering permanent ill effects.*

NATIONAL AERONAUTICS AND SPACE ADMINISTRATION

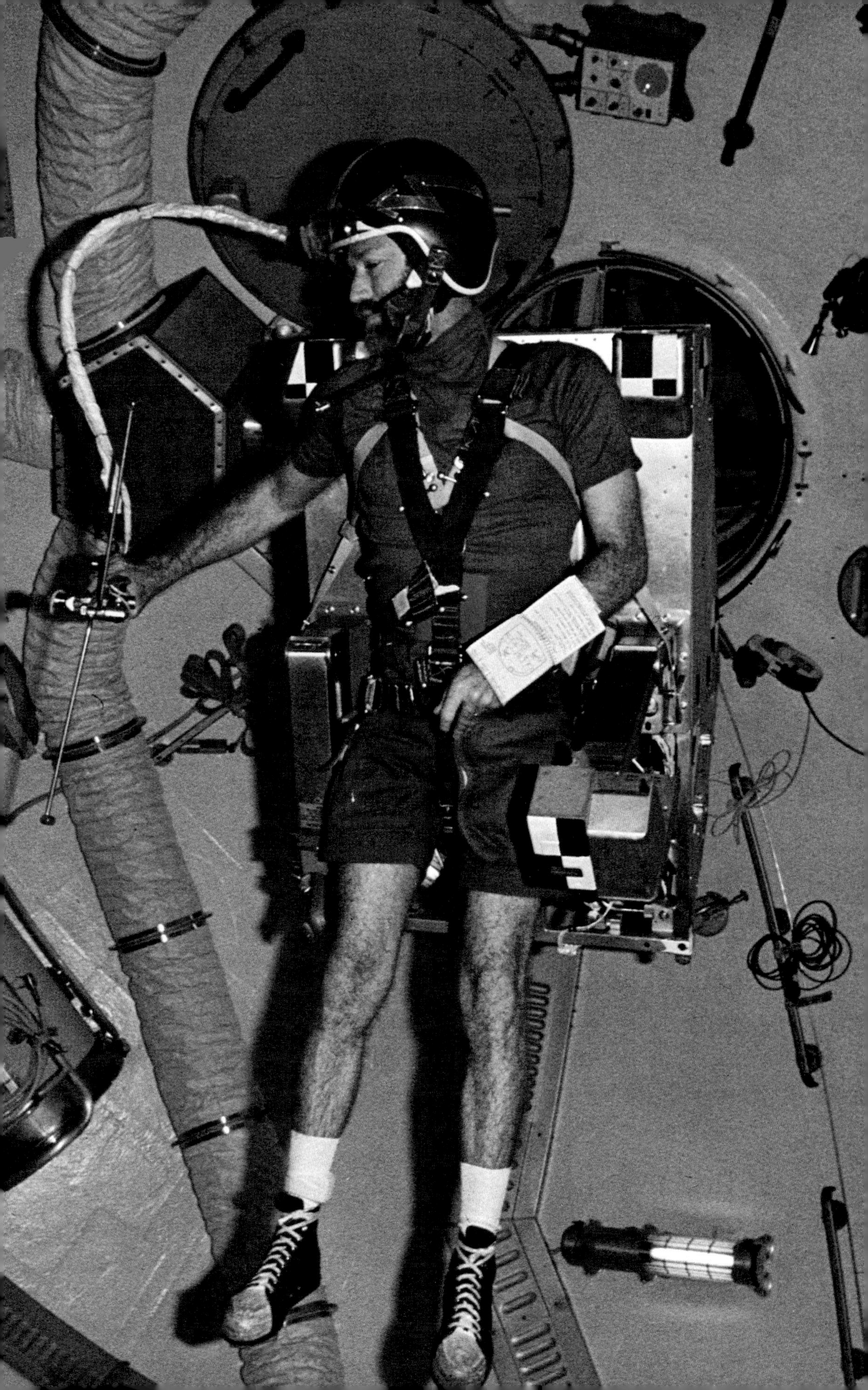

5/ VERSATILE FLYING ROBOTS WORK TO BENEFIT MAN

As I edged into the rush-hour stream of Los Angeles freeway traffic and turned east toward Pasadena, hundreds of automobiles raced past me in both directions, carrying their occupants to work. Here on earth it seemed a very ordinary day. Yet I knew that at that moment, thousands of miles out in space, something most extraordinary was taking place: Other man-made vehicles were crossing new bars and soundlessly penetrating vast new oceans.

Two Russian spacecraft, Mars 2 and Mars 3, were on a 287-million-mile journey to the red planet, while slowly overtaking them was the smaller United States craft Mariner 9.

Today I was to talk with some of the men who had fashioned Mariner's electronic brain and programmed it to send back long-concealed Martian secrets; and I was to watch them prepare to send across the ever-widening gulf of the cosmos the commands that would keep their charge performing as planned.

At the base of the San Gabriel Mountains I came to the ultramodern buildings of the Jet Propulsion Laboratory, an affiliate of California Institute of Technology. In the lobby, glass cases of medals, plaques, and citations testified to the achievements of JPL and its distinguished director, Dr. William H. Pickering.

The newest planetary probe was now in the third day of its 5½-month voyage. On a sign in the lobby that had formerly read, "Mariner '71 GO," hand-drawn letters had been added. The new version: "Mariner '71 GONE! Congratulations!!"

"Who changed the sign?" I asked a secretary.

"No one knows," she said. "We found it when we got here this morning."

The mood of relief and confidence was evident in Bill Pickering's boyish smile and firm handshake. His manner was that of a new father whose expectations had been gratifyingly confirmed.

"Of course, we're happy," he said. "This is not just another probe; Mariner 9 is the first of a series that we hope will return data on all eight planets

OFF TO SCOUT *the Van Allen radiation zone—in a prelude to manned missions—Explorer 6 leaves Cape Canaveral aboard a Thor-Able rocket, August 1959. This "paddlewheel satellite" deployed four paddle-shaped test vanes of solar cells that successfully converted sunlight to electric power.*

UNITED STATES AIR FORCE

within the next ten years." Dr. Pickering echoed the hope already expressed in Washington by NASA Deputy Administrator George Low that the Soviet probes would also get good data, and that the two countries would agree to exchange information—an agreement that was in fact concluded officially several weeks before any of the spacecraft reached Mars.

Near the end of our discussion my eyes wandered to a large orange-and-black drawing hanging on the office wall. It looked almost like a child's primitive impression of a planet.

Dr. Pickering smiled. "That was our very first view of Mars from Mariner 4 on July 13, 1965. Our people couldn't wait for the computers to turn the data into a complete pictorial presentation, so they did this one by hand and presented it to me."

I found it reassuring that human minds and hands still found expression in this citadel of automation. As information officer Frank E. Bristow led me into the control center, the arrangement of consoles and vividly colored displays reminded

ECHO 2 *communications satellite balloons enormously inside a dirigible hangar at Lakehurst, New Jersey, in a prelaunch test. Boosted aloft in 1964, the plastic sphere inflated when residual air inside expanded in the vacuum of space and a powder called pyrazole, warmed by the sun's heat, turned to gas. The satellite, 135 feet across, reflected radio signals beamed from earth. Below, a backup balloon for 1960's Echo 1 lies in a cutaway capsule at the Smithsonian Institution, Washington, D. C.*

N.G.S. PHOTOGRAPHER JAMES E. RUSSELL (BELOW); NASA

DONALD J. CRUMP, N.G.S. STAFF

VIA SATELLITE *becomes a byword in nation after nation as "talking moons" enter the heavens. In 1962 the 170-pound experimental Telstar I (below) pioneered amplified-signal relay satellites. By 1974 more than 80 countries belonged to the International Telecommunications Satellite Organization (Intelsat), managed by the U. S. company, Comsat. At right, an Intelsat IV undergoes integrity checks in an anechoic chamber at a Hughes Aircraft Company plant in California. Poised over the Pacific Ocean in February 1972, the 1,587-pound satellite brought to U. S. living rooms television coverage of conferences between Premier Chou En-lai and President Richard M. Nixon during his visit to the People's Republic of China.*

AMERICAN TELEPHONE AND TELEGRAPH

me of the command room of the spaceship *Enterprise* in the "Star Trek" television series. I almost expected the door to slide open by itself with a soft swish; instead, Bristow opened it by inserting a coded plastic card.

From here in the Space Flight Operations Facility, man can precisely execute his will millions of miles away.

In the darkened room, jagged red flashes on a huge screen showed that Mariner's signals were coming in from the Goldstone tracking station in California. The youthful-looking chief of Mariner operations, Tom Hayes, pointed to a closed-circuit TV screen on which a confusing array of letters and figures was constantly changing. "That's Mariner's heartbeat," Hayes said. "You can see the nitrogen jets, or thrusters, giving small kicks."

"Where?" I asked, bewildered.

Hayes put his finger on one of the numbers. "Let me translate for you. See this P GYRO/SS jump to 54? That shows the yaw jet kicking to hold the alignment with the star Canopus and the sun. See this, CT INTENS 71? That shows Canopus's intensity is now 71, just where it ought to be in the field of view. See this 219? It just went to 220. That shows the spacecraft is drawing 220 watts through its solar panels. All of this data helps us verify the proper operation of the craft in preparation for mid-course correction."

I asked how he sent a command to Mariner, and Hayes led me to a small gray box with a typewriter-like keyboard.

"This is called a 2260 logical unit. You enter the coded command here on the keys, then to send it you hit this black space bar—just like that."

Hayes explained that on that day, June 1, 1971, Mariner was only 360,000 miles away. "Right now the signal, traveling at the speed of light, takes just under two seconds to get to the spacecraft. Mariner will be more than 75 million miles away when it gets to Mars, and a command signal will take *seven minutes* to get there."

When Mariner 9 reached Mars on November 13, the first pictures suggested the mission might be futile. The entire surface of the planet was being lashed by a huge planetary windstorm. Dust blown at speeds estimated in excess of 200 miles per hour obliterated the landscape. For nearly two months Mariner 9 circled and waited. Then, miraculously, the great storm abated.

As the skies cleared, Mariner's instruments

PHOTOGRAPH COURTESY HUGHES AIRCRAFT COMPANY

began what Dr. James C. Fletcher, NASA's administrator, later called "one of the great scientific and technical achievements of this decade." In the words of JPL's Frank Bristow, "the long weeks of frustration ended with one magnificent and exciting word—VOLCANOES!"

More than 1,500 mapping photographs permitted the reconstruction of a fascinating world of gigantic volcanoes, one with a diameter of up to 370 miles, whose peak towers 15 miles above a plain; massive lava flows; wind-scoured chasms that could swallow several Grand Canyons; features that showed apparent erosion by water; and regions that changed character before Mariner's eyes. In one case, a dark arrowhead-shaped area seven miles across appeared in just 13 days as layers of loose, light-colored material were blown away. Beneath thin white clouds, the surface temperature of Mars varied in Mariner 9's instruments from −150° F. on the night side to 70° at noon on the planet's equator.

In the history of exploration there had never before been anything like these American and Russian robots, created not in man's image but reflecting his profound curiosities and aspirations.

Since ancient times the planets have beckoned, and as early in the Space Age as February 1961 the Soviets ambitiously sent Venera 1 toward Venus. When radio communication was lost after the pioneer probe had traveled 4.5 million miles, the Russians took an unusual step. They quietly sent a delegation to England's famous radio observatory at Jodrell Bank to see if contact could be regained. Unfortunately it could not. Somewhere in the terrible bake-freeze of distant space, Venera 1 flew on, forever mute.

As the Soviets experienced additional difficulties in shooting for Venus and Mars, the United States succeeded with two of its first four planetary probes. In 1962 Mariner 2, after traveling for 109 days, swept within 21,000 miles of Venus.

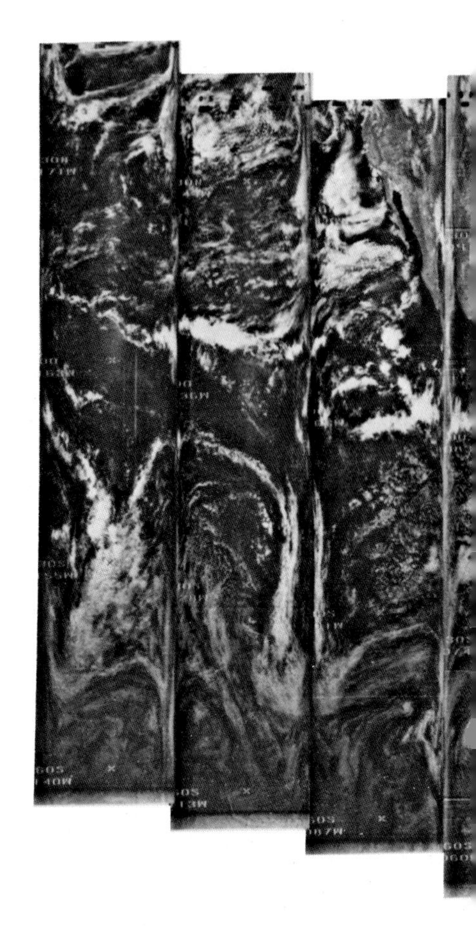

NATIONAL AERONAUTICS AND SPACE ADMINISTRATION

WEATHER WATCH: *Strip-by-strip views make up a global panorama, upper right, that helps meteorologists determine the forces that shape earth's weather patterns. Nimbus 3 transmitted the photographs during daylight portions of its polar orbit. At right, a meteorologist at the National Oceanic and Atmospheric Administration in Suitland, Maryland, prepares a weather map from Nimbus pictures taken from an altitude of 600 miles.*

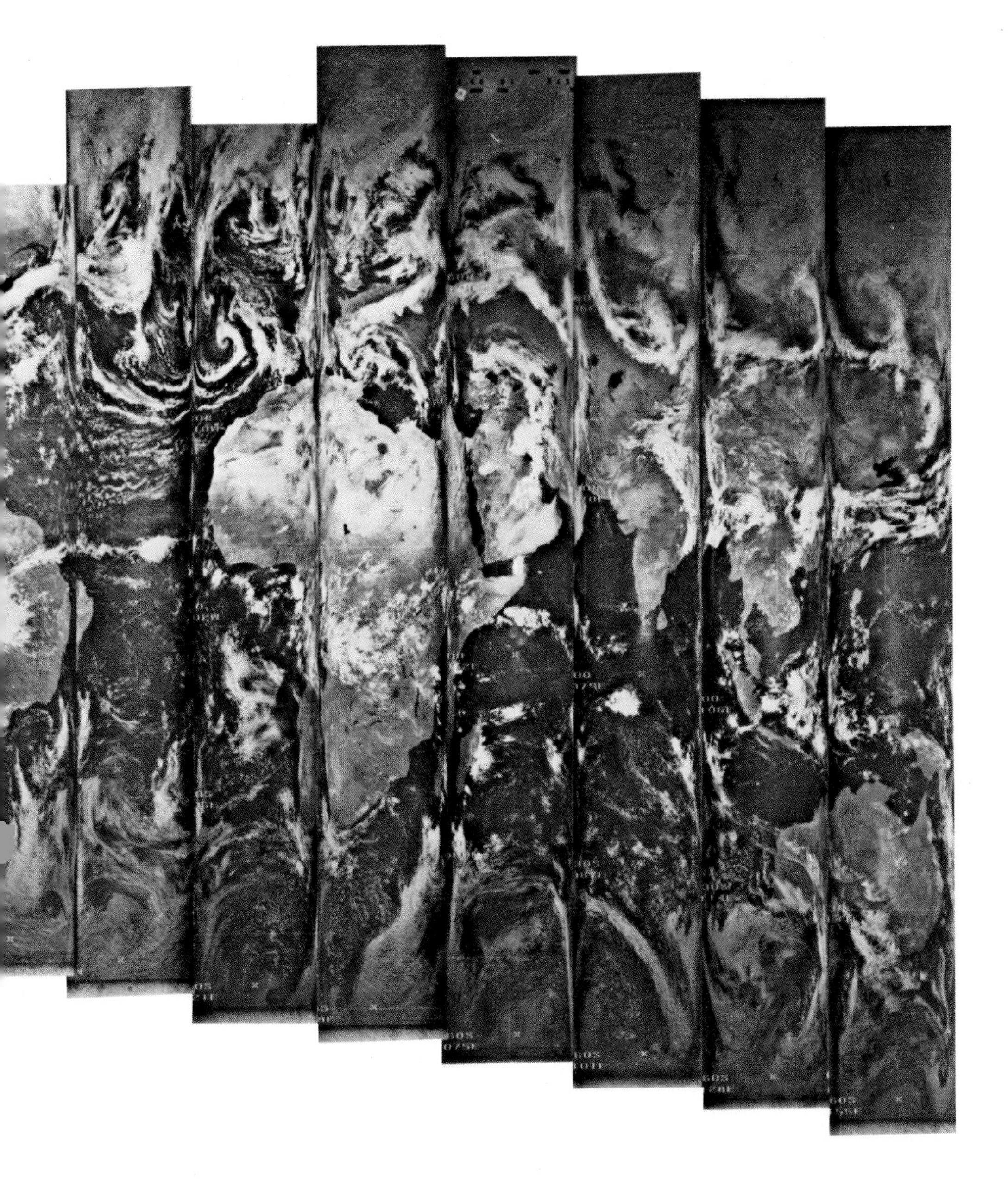

What its instruments reported was a blow to those who had hoped the thick atmosphere might conceal some form of life. Instead, the atmosphere appeared to have an extreme greenhouse effect, smothering the surface with heat of 750° F.

The spectacularly successful Mariner 4 transmitted equally startling news about Mars. After a 228-day journey covering 350 million miles, the craft, which resembled a big jeweled windmill, approached the red orb in mid-July 1965. Within 23 hours picture data, in the form of numbers convertible to shades of light and dark, began to come in at the Jet Propulsion Laboratory. During the next several days a computer translated the electronic language into photographs.

"When we hit frame 7, it was a very dramatic moment," recalls Dr. Bruce C. Murray of California Institute of Technology. "We began to recognize many, many craters. I don't think very many people had expected Mars would look like this."

At a crowded press conference, reporters insistently sought an answer to a centuries-old question: Was there any evidence of life on Mars? No, the scientists replied, but we can't expect that kind of detail from a few Mariner pictures. Even earth, photographed thousands of times from meteorological satellites, rarely reveals any evidence of intelligent life in those pictures.

After hitting Venus once but with minimal data return, the Soviets finally had their V-for-Venus day on October 18, 1967. Venera 4, still in radio contact with earth, arrived near our brightest planet after a four-month trip. It ejected a 40-inch egg-shaped package that began to descend through the alien atmosphere. The Venera spacecraft burned in the intense heat of entry, but the heavily insulated capsule drifted down by parachute.

As the instrument package was tossed by winds of greater than hurricane force and broiled by intense heat, it took a wide range of readings. For 96 minutes, it radioed to earth a story of the hostile Venusian atmosphere. Cloud temperatures: 104° to 536° F. Atmosphere: 98.5 percent carbon dioxide, no nitrogen.

The new data indicated that Venus is, indeed, a "hot planet," one that would be difficult for man ever to visit. "It would be like wandering around the depths of a boiling ocean in a submarine," Dr. Pickering of JPL told me.

The Soviets nevertheless continued their dogged

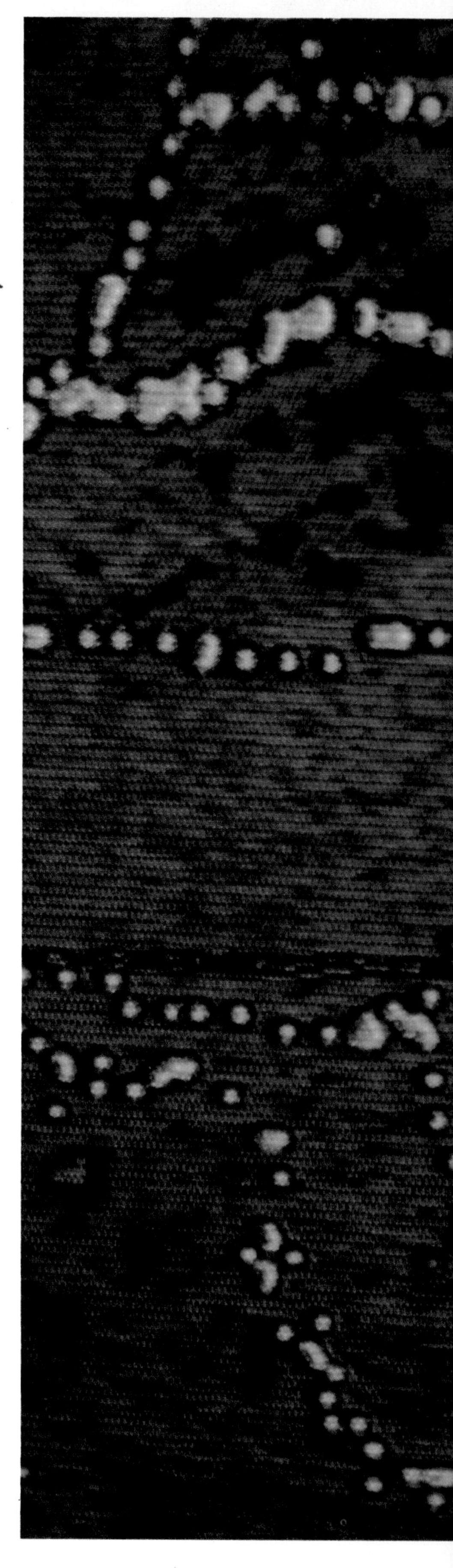

HOUSES SHREDDED *to lumber, shrimpboats driven into buildings—in the debris of 1969's Hurricane Camille, a vessel lies 300 yards from subsiding waters near Gulfport, Mississippi. Hurricane Ginger, long-lived storm of 1971, blotches the Atlantic Coast on a color-scanning densitometer, a device that converts black-and-white satellite pictures into hues where thickest clouds show brightest. Such images guide forecasters' warnings; early notice saved many lives from Camille.*

N.G.S. PHOTOGRAPHER OTIS IMBODEN (BELOW); G. E. ARNOLD

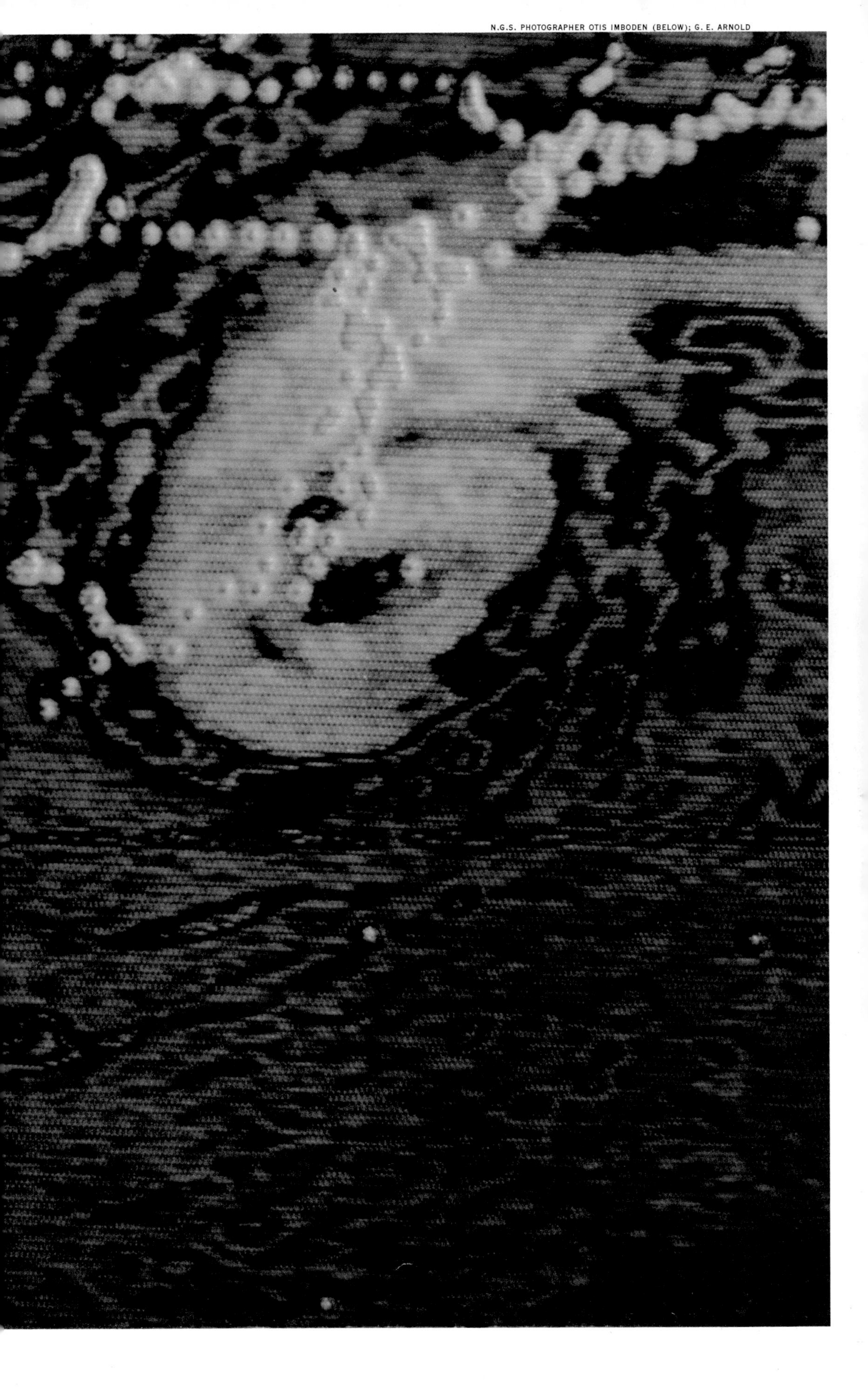

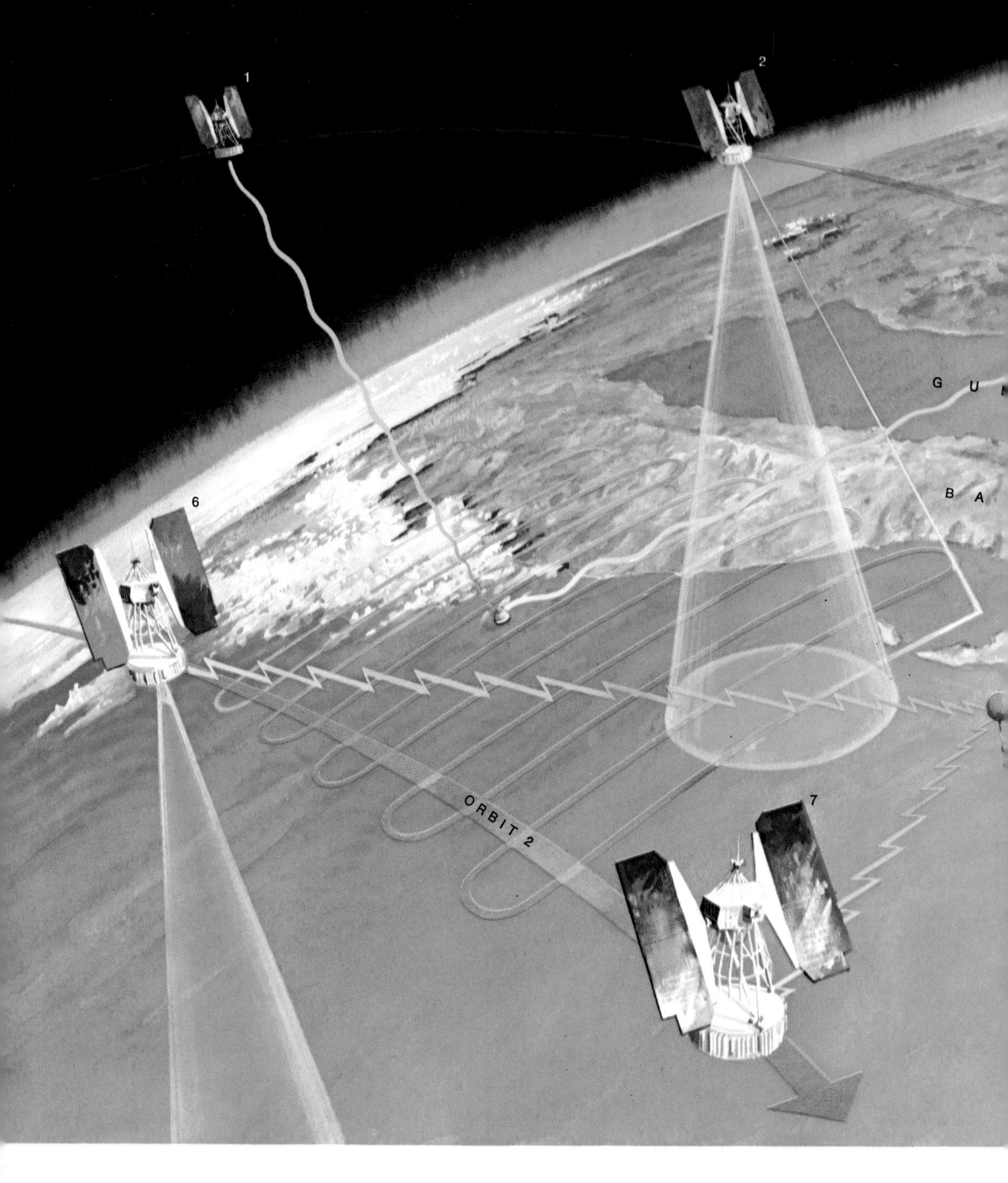

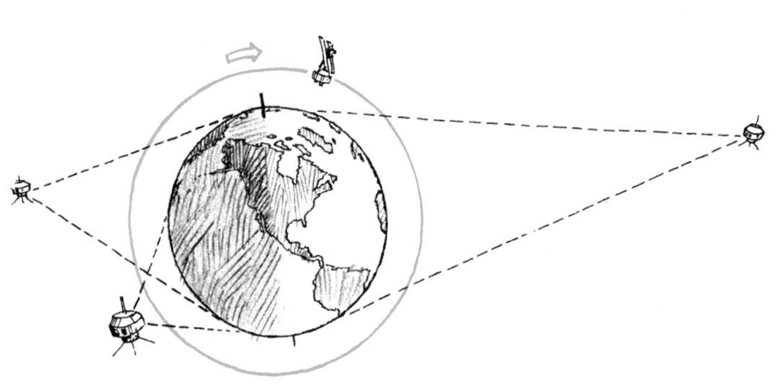

WEATHER SATELLITES *in low polar and high equatorial orbits (left) will blanket the globe in the 1970's. A satellite 600 miles high orbits 14 times from pole to pole as the earth completes one revolution; in 12 hours, all the earth's surface moves beneath it. Synchronous satellites 22,300 miles above the Equator will keep constant weather watch. Dotted lines indicate the area monitored by each orbiter.*

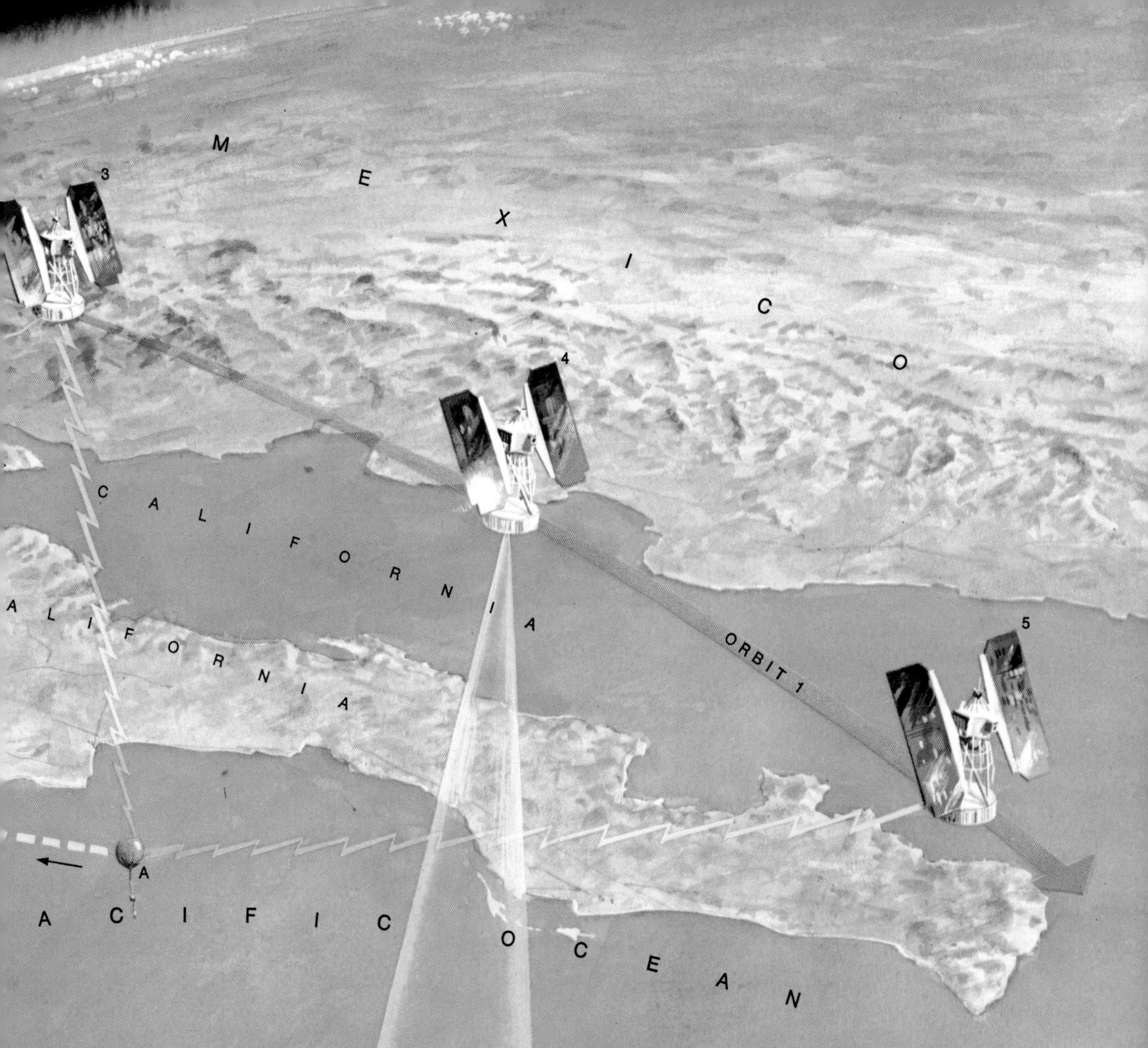

Meteorological satellite of the 1970's gathers data from a network of ground and air stations. On a pass over Baja California it activates telemetry equipment in a buoy (1) and continuously probes earthward (2) with infrared and microwave equipment to measure temperature and humidity at varying altitudes, and to scan cloud cover. After collecting information from the buoy on air pressure and ocean temperature (3), the satellite makes contact with a weather balloon (A). It continues to probe the atmosphere (4) and triangulates to determine the position of the balloon (5). By the time the satellite returns for another pass (6-7), the earth's rotation has exposed a new area to surveillance. Further triangulation of the balloon's new position (B) indicates wind velocity and direction.

PAINTING BY RICHARD SCHLECHT

investigation with Veneras 5 and 6 in 1969. The Venera 6 capsule sent back a great deal of information during its parachute descent, until it was crushed when the outside pressure exceeded 27 times that of earth's atmosphere.

Two U. S. Mariner probes in 1969 were designed to send back data, including TV impressions of the Martian surface, that might prove useful in planning future missions to seek evidence of life on the red planet. Both Mariners 6 and 7 sent back clear images of the surface, transmitting 64 shades of light and dark through the language of their computers. The 200 pictures confirmed a desolate appearance in the areas viewed by the cameras. Other data offered only a remote possibility that any form of life could exist on the planet.

After weeks of analysis, scientists speculated that the south pole of Mars consists in part of dry ice—frozen carbon dioxide—possibly mixed with ordinary ice. The report pointed up the apparent scarcity of water and the probability that nitrogen is missing, concluding that "if this is true, any life chemistry on Mars will have to be very much different" from what we know at present.

In December 1973 Pioneer 10 became the first man-made object to approach Jupiter. Just to get there, it had to survive the asteroid belt, pass through Jupiter's own "bow waves" caused by the impact between the planet's magnetic field and the solar wind, and, finally, withstand Jupiter's fearsome radiation belts. Early analysis of Pioneer's data showed Jupiter to have an atmosphere containing hydrogen and helium. Jupiter radiates more than twice as much heat as it receives from the sun. The planet's mysterious reddish spot, the data indicated, may be a giant mass of raging and swirling russet clouds towering as much as three miles higher than the planet's other clouds. Intermingled with the lower clouds are bands of color that race at vastly different speeds around the massive planet. Pioneer 10 confirmed that Jupiter's magnetic field is backward by earth standards; a magnetic compass on Jupiter would point south. The planet was so full of new mystery that it was clear that earthmen would be preoccupied with it for decades to come.

Using the impetus of Jupiter's powerful gravity, Pioneer 10 sped on to an unknown fate, the first object from earth destined to escape the solar system. In case it should ever reach a star system with beings capable of intercepting it, NASA included drawings and symbols in an attempt to give some indication of the location of its home planet and of the beings that inhabit it.

Pioneer 11 was launched in April 1973 for its journey to Jupiter and Saturn.

That Venus is inimical to life as we know it was further confirmed by the Soviet Union with the 1970 flight of Venera 7. This time the formidably insulated capsule drifted all the way to the superheated surface, and broadcast for 23 minutes after it landed. The new data were substantial—and not very inviting. Who would want to visit a planet with a surface temperature that measured 854.6° F., more than hot enough to melt lead, and a surface atmospheric pressure 90 times that of the earth at sea level?

In July 1972 the Soviets landed Venera 8 on the planet's scalding surface and obtained data from it for nearly an hour. The robot was an ingeniously pre-chilled, egg-shaped probe, built almost as strong as a cannonball. Before its signal stopped, Venera 8 relayed some significant and slightly different new data on Venus: surface temperature of 880° F.; surface pressure about 100 times that on earth, and surface soil density about half that of earth; some of the same radioactive soil elements as in many volcanic rocks on earth, indicating that Venus—like earth, Mars, and earth's moon—had once been hot enough to melt; high-altitude winds similar to earth's jet streams but—unlike Mars—surprisingly light surface winds. The data also confirmed that because of Venus's greenhouse effect, the surface does not cool during its months-long nights.

Soon after Pioneer 10 reached Jupiter, Mariner 10, which was scheduled to fly past both Venus and Mercury, focused its instruments on the cloud-veiled Venusian globe. Black-and-white images taken through ultraviolet filters by Mariner's television cameras revealed a turbulent rise of hot gases from the planet's equatorial zone. At least part of the gases apparently rose through a strangely migrating dark spot, or hole. Then the gases expelled by the equatorial heat engine spiraled toward both poles to cool and descend.

Mariner 10 found small amounts of hydrogen, oxygen, helium, and argon in an atmosphere that is almost entirely carbon dioxide. It also detected carbon dioxide being spewed into the space immediately surrounding Venus. The craft was the first to use a planet's gravity to slow itself down so it could intercept tiny, sun-scorched Mercury.

In March 1974 Mariner 10 became the first two-planet probe to reach both its targets. The close-up pictures showed a small planet that seemed at first to resemble the moon. Mercury was blasted with random, overlapping craters; in places there appeared to be level plains. But after more pictures were examined—some from as close as 3,212 miles—Dr. Donald Gault of the Ames Research Center in California said, "There's something definitely different down there in the structures and patterns." The data also indicated that Mercury has an extremely thin atmosphere apparently containing only helium; a magnetic field, and possibly lava flows.

Just as robots were our first emissaries to the planets, so had they earlier led the way to earth's own satellite. Neither the Soviet Union nor the United States, once it had the necessary rocket vehicles, lost any time in sending automated explorers on missions to the moon. The first 13 U. S. shots failed, despite my personal encouragement. Once, for example, I persuaded a technician to stencil near the rocket's tip a poetic pat from Tennyson: "After it, follow it, follow the gleam!"

The Russians, on the other hand, scored in September 1959, when Luna 2 struck the moon. Three weeks later, Luna 3 sent back the first photographs of the far side of the moon.

Ranger 4 hit the moon in April 1962, but returned no data. Success came for the U. S. in July 1964, when Ranger 7 took 4,316 still pictures before crashing in the Sea of Clouds.

Scientists could send a probe to crash into the moon, but could they put one there intact? Years of delay for the U. S. and failure for the Soviet Union plagued efforts at a soft-landing. Finally in February 1966 a 220-pound Soviet automated station, round as a beach ball, touched lunar soil.

During the first minutes after impact, no one knew whether it had withstood the landing. Then four petal-like segments of the outer shell folded down slowly, forcing a TV scanner into topmost position and allowing four antennas to extend. Fifteen minutes after landing, the first moon station began sending signals to earth.

Five U. S. and two Soviet instrument packages, including mechanical probers and diggers, dropped softly onto the moon during 1966 and 1967. Their accomplishments and those of five Lunar Orbiters established with certainty that man could go to the moon.

In 1968 the Soviets sent around the moon Zonds 5 and 6, containing a variety of life forms. The missions not only involved important biological experiments but also were the first attempts to bring spacecraft back through the earth's atmosphere at what the Russians call "the second cosmic speed"—approximately 24,500 miles per hour, equivalent to the velocity an outward-bound craft must attain to escape earth orbit. (The "first cosmic speed," about 17,500 miles per hour, is the rate necessary to achieve earth orbit.) Difficulties with the Zond 5 re-entry caused deviations from the flight plan, but the Zond 6 return was completely successful.

In July 1969, even as Neil Armstrong and Buzz Aldrin were preparing to descend to the Sea of Tranquillity, the Luna 15 Soviet automatic probe was also in lunar orbit; but it was to crash in the Sea of Crises just a few hours after the Apollo pioneers landed 500 miles away. Perhaps Luna 15 was attempting to do what Luna 16 accomplished in September 1970, when it soft-landed in the

JOHN CRAIGHEAD

FAUNA—AND MAN—*benefit from weather satellites. Photographs of Baltic Sea winter conditions transmitted from space help direct the Finnish icebreaker* Tarmo *to harbors that need clearing. In Wyoming, an elk—the first free-roaming animal tracked by satellite—wears an electronic collar that links it with Nimbus 3. The use of space apparatus to monitor wildlife in this and similar experiments promises new conservation techniques.*

N.G.S. PHOTOGRAPHER GEORGE F. MOBLEY

FAIRCHILD INDUSTRIES

PREVIEW FOR A CLASSROOM: *Near New Delhi, Indian families meet at a village TV set; a rooftop dish antenna made of chicken wire brings in the signal. Thus, if all goes well, Indian teachers in 1975 will transmit educational programs via the world's most advanced satellite; America's ATS-F (left, in full-scale mock-up). Made available to India for a year, ATS-F with its powerful signal will serve village receivers costing only a few hundred dollars each. Applications Technology Satellites, hovering in orbit 22,300 miles high, can reach the most remote countries in the world.*

Sea of Fertility. Commands from earth activated a tool that obtained both loose lunar material and a core sample. These were drawn inside a short, squat rocket atop Luna 16; then another signal blasted the rocket aloft as the first step in its remote-controlled journey back home.

The quantity of material returned to eight Soviet laboratories was only 3.56 ounces, but the method represented a fundamental shift of emphasis in Soviet employment of automated devices. "Luna 16," said Academician Georgi Petrov, "played the role of an active operator functioning on commands from earth. The future belongs to such automatic explorers."

Two months later the Soviets tested a truly remarkable "active operator" with Luna 17. Three hours after the automated station landed in the Sea of Rains, five Soviet ground operators working in concert prepared to deploy from it an amazing machine called Lunokhod 1. As the ground controller observed both panoramic and stereoscopic pictures televised by Lunokhod, the eight-wheeled vehicle was directed down a ramp. Each wheel was powered by its own electric motor.

The Soviet purpose was not just to track-test the Lunokhod. During the first five days, as it was directed along an intricate course of more than 180 yards—over ridges, across rilles, and up and down crater walls—its electronic brains were working full time. On board, in addition to communications equipment, were an X-ray telescope to measure radiation from beyond our galaxy, an X-ray spectrometer to determine the composition of lunar rocks, a penetrometer to measure density of the lunar surface material, and a French-made device to reflect laser beams. By the end of December the robot Lunokhod had traveled almost a mile. Before it stopped functioning in September 1971, it had covered 6.5 miles.

After Luna 18 crashed during an attempted landing in September 1971, Luna 19 achieved an intended lunar-photography orbit a month later. Luna 20 arrived on February 18, 1972, and landed three days later in the rugged Apollonius Mountain Range. On February 25, new lunar samples from this probe were successfully soft-landed in the Soviet Union.

Still another Soviet automatic station, Luna 21, in early 1973 released Lunokhod 2 in a crater named Le Monnier, about a hundred miles from the Taurus-Littrow site of Apollo 17. The bathtub-shaped Lunokhod 2, which weighed 1,848 pounds, discovered soil with a lower iron content than had Lunokhod 1.

Will Lunokhod's ground mobility someday be combined with the return capability of Luna 16 and 20 to explore Mars and other planets? We would be foolish now to doubt such possibilities.

Despite the already proven capabilities of automated planetary and lunar explorers, the most immediate and beneficial returns by far from both Soviet and American space robots have been

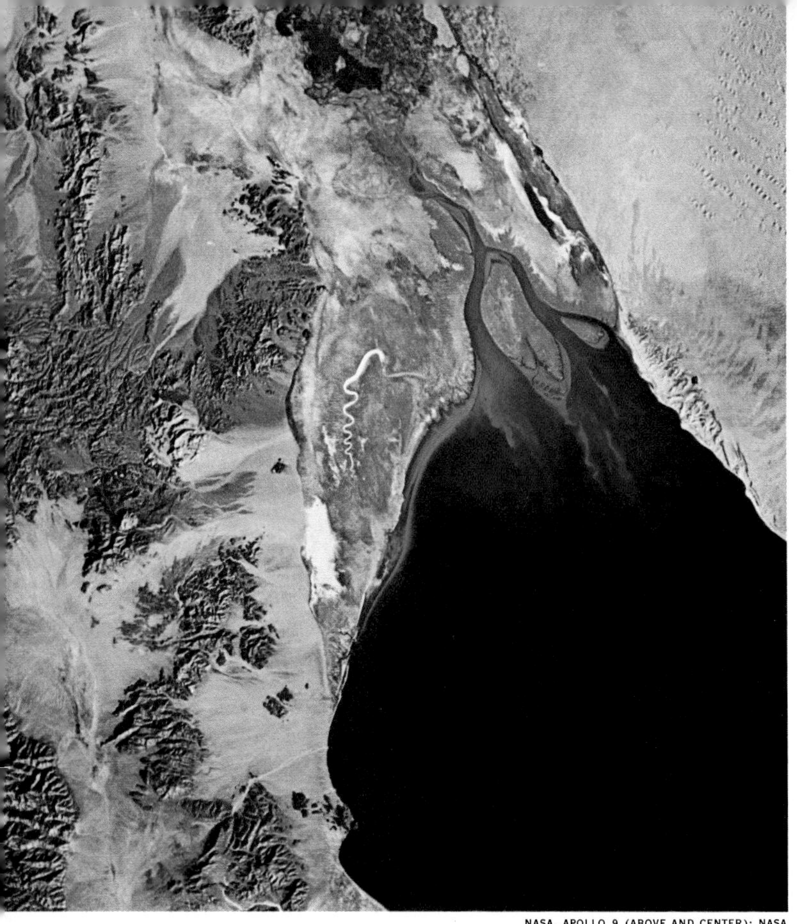

NASA, APOLLO 9 (ABOVE AND CENTER); NASA

those yielded by the hundreds of unmanned vehicles orbiting the earth.

Dr. John E. Naugle, then NASA's associate administrator for the Office of Space Science and Applications, says that more than 1,200 packages of instruments "have faced toward earth instead of outward toward space. They have provided us with long-distance communications, advance weather information, navigation guidance, and unique views of the earth's surface.

"These automated craft have had truly dramatic impact on the world we live in. And vast new wealth may well be derived from growing industries based on satellite technology and from natural resources pinpointed by remote sensors. When you realize how rapidly instruments and spacecraft are growing in sophistication, you know that we have just begun."

A sketch of the packages of electronic instruments whirling around the earth would resemble a diagram of electrons flying about the nucleus of a complex atom. Each package doggedly follows its set path north-south, or west-east, or somewhere in between. Each orbiter travels at a speed dictated by its distance from earth—faster if close to earth, slower if far away. The shape of the orbit may be round, like that of the Early Bird communications satellite, or extremely elliptical like those of Soviet communications satellites.

Into the newly annexed 60,000-mile extension of earth's realm, man rocketed about 1,500 satellites of all kinds during the first 16 years of the Space Age.

Every day spacecraft relay thousands of transoceanic conversations, and feed billions of electronic bits of information onto magnetic recording tape. Two U. S. processing stations—Goddard Space Flight Center in Maryland and the Jet Propulsion Laboratory in California—receive thousands of miles of such tape every year.

From the beginning, neither the U. S. nor the U.S.S.R. wasted time rocketing empty boxes into space simply to prove that an object could be put in orbit. Sputnik 1 beeped its way into the memory of people throughout the world as its transmitter sounded the ionosphere and made indirect measurements of space temperatures. Minutes after launch, our tube-shaped Explorer 1 reported

INVENTORY FROM ORBIT: *To help find means of supporting an increasing population, NASA's Earth Resources Technology Satellite program assesses the planet's limited capacities. Photographs from space indicate the potential for ERTS multiple sensors. An infrared photograph of the arid Colorado River delta reveals trees and other foliage as red patches above the estuary. Snow depths in the Rockies foretell floods—or gentle runoff to nourish food crops; on the Queensland coast of Australia, reefs show as clearly as smoking canefield fires.*

higher radiation counts above 600 miles than anyone had suspected.

In the first 15 months of the Space Age, the U. S. and the Soviet Union had sent up at least one package of instruments for nearly every kind of task the satellites perform today. By the early 1970's, engineers in both countries had created several generations of satellites, from simple short-lived workers to highly complex and long-lived professionals with brilliant computer brains.

An early task for satellites—that of communication between distant points on earth—was suggested as early as 1945 by Arthur C. Clarke. Thirteen years later, Score went up hundreds of miles and broadcast President Eisenhower's Christmas message to the world.

In August 1960 the United States launched one of the most famous objects ever to circle the earth. Echo 1, a silvery, thin mylar balloon 100 feet in diameter, orbited the earth every 114 minutes.

Many other communications satellites, or satcoms, some of whose names became as familiar as Echo's, soon followed. One was a 30-inch sphere called Telstar, owned by the American Telephone and Telegraph Company and launched in July 1962. As the first privately owned satellite reached a point in its orbit that brought the coasts of the United States and France within view, AT&T engineers began transmissions from Andover, Maine. A fraction of a second later, a picture of the American flag rippling in a gentle breeze appeared on a television screen at the Pleumeur-Bodou receiving station more than 4,200 miles away.

Recognizing the importance of worldwide communication, Congress in 1962 authorized establishment of the Communications Satellite Corporation—Comsat. By 1974 the publicly owned corporation had drawn more than 80 nations into

1
2
4
5
7
8

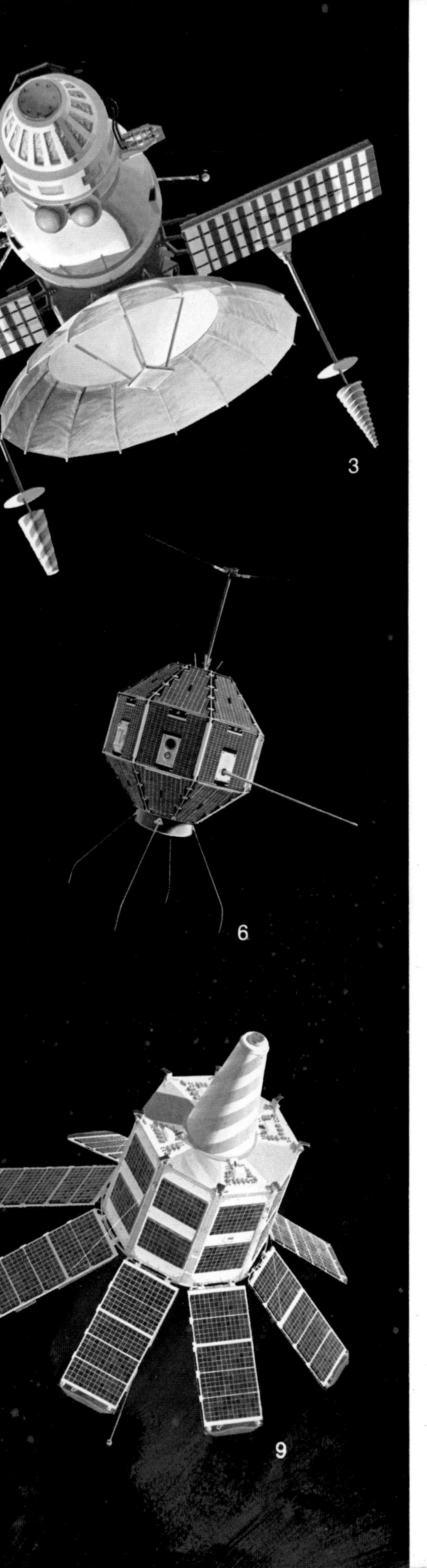

a network called the International Telecommunications Satellite Organization—Intelsat.

In the spring of 1965 the consortium invested in the "synchronous satellite"—a high-altitude satellite that appears to stand still by keeping pace with earth's rotation. Soon after launch on April 6, Intelsat 1 (Early Bird) proved a remarkable success. "Overnight... [it] ... increased transatlantic communications capacity by 50 percent," James McCormick, then Comsat chairman, told me.

The same month Early Bird was launched, the Soviets announced—on April 23, 1965—their own communications satellite, the Molniya, or Lightning; twenty were aloft by the end of 1971.

Since communications satellites were developed, worldwide television audiences have been able to witness events as varied as President Nixon's 1972 visit to China, the wedding of Britain's Princess Anne, and the launching and recovery of manned space-flight crews. In April 1974, Westar I opened a new era of transcontinental communications. The Western Union "stationary" satellite, which can handle eight million words per second, is able to relay voice communications, as well as Telex, TWX, and TV, between any two points in the United States.

The U. S. military put up its first 26 communications satellites using economical new techniques: assembly-line production of identical

FOREIGN-BUILT SATELLITES—*many launched by NASA—take national skills into space, adding to man's knowledge of the cosmos. (1) Canadian Alouette 1, fired aloft in 1962, transmitted data on electron density for a decade. (2) The Astronomical Netherlands Satellite, due for launch during 1974, will monitor ultraviolet radiation and X-rays. (3) The U.S.S.R. sent Veneras 5 and 6 to probe the dense atmosphere of Venus in 1969. (4) Britain's Ariel 4 made low-energy particle and radiation studies. (5) Italy's San Marco 3 rose from a platform off the Kenya coast in 1967 to radio data on the equatorial atmosphere. (6) Japan launched Shinsei in 1971 to measure cosmic rays and other phenomena. (7) In the most ambitious joint effort yet, the United States and West Germany plan Helios as a solar probe for 1974-1975. (8) The HEOS, for Highly Eccentric Orbit Satellite, works in interplanetary physics for ESRO, the European Space Research Organization. (9) Constructed by the French, Eole first measured winds, then kept track of icebergs, now monitors buoys for NASA.*

PAINTINGS BY FRANCIS J. KRASYK (NUMBER 1 EXECUTED FOR NASA)

PLANETARY PROBES REACH DISTANT WORLDS

Extending man's knowledge of the solar system, bold space probes provide new views of the planets. A massive ice cap covers the southern polar region of Mars (left) in a picture transmitted in 1969 by Mariner 7 while 337,132 miles from its surface. Frozen carbon dioxide—dry ice—and ordinary ice make up this cap and another in the northern hemisphere. The polar masses expand and contract with the changing seasons: As spring arrives in one hemisphere, its cap begins to evaporate, while the cap at the opposite pole grows. The Venusian cloud cover (center) shows up as a swirling pattern in close-ups transmitted from Mariner 10. Scientists confirmed from the flyby data that the 50-mile-thick atmosphere contains at least three major layers; it has poisonous gases and some oxygen; and its convection currents resemble earth's atmospheric movements. Mariner 10 transmitted this view of Venus from a distance of 450,000 miles on February 6, 1974, while headed for a rendezvous with Mercury. Many craters pock Mercury's moonlike surface (upper right) in this March 28, 1974, television picture taken nearly 600,000 miles from the planet. Some Mariner project scientists, however, believe Mercury's interior may resemble the earth's more than

100-pound models shaped like tops; the launch of more than one satellite at a time with one rocket; repair by remote control.

The first set of seven went up in June 1966, to be kicked one by one out of the carrier. The final eight were dropped above the Equator like beads on a string in June 1968. Each satellite reports to earth how its working parts are behaving, and each has a section of "spare parts."

Recording images of the earth from space has become a highly developed specialty of the military. Satellites employing a technique known as "folded optics" (bouncing light between mirrors inside a camera before it reaches the film) can identify objects as small as a compact car.

Numerous U. S. satellites use high-resolution cameras and a growing array of remote-sensing instruments. The Russians have done the same with more than 260 of their Cosmos satellites.

During the India-Pakistan War in December 1971, the Soviets were believed by U. S. authorities to have dispatched two observation satellites over the area for a firsthand look at the military situation. The United States is at work on similar "quick-look" satellites to help avoid miscalculations caused by tardy or inaccurate information.

that of the moon. Jupiter's giant red spot looms above bands of whirling clouds (right) in a picture relayed by Pioneer 10 on December 1, 1973, from a distance of 1,580,000 miles. The Jovian moon Io casts a dotlike shadow on the huge planet. Data from the flyby suggest that a turbulent mass of clouds accounts for the red spot. On December 3, 1973, Pioneer 10 swept within 81,000 miles of the cloudtops in what NASA Administrator James Fletcher called "an engineer's dream come true." After passing by Jupiter, Pioneer 10 continued its outward flight, destined to become the first man-made object ever to leave the solar system.

NATIONAL AERONAUTICS AND SPACE ADMINISTRATION

Scientists predict that military and scientific satellites will greatly affect activities ranging from map making to prospecting, from fishing to farming. The early Nimbus, for example, with its infrared camera for seeing clouds at night, also spotted big icebergs, traced Gulf Stream temperature patterns, and discovered underground freshwater rivers emerging in bays and oceans.

Remote-sensing devices assist our national defense in various ways. Vela satellites, launched in pairs into an orbit 70,000 miles high, detect evidence of nuclear detonations, a monitoring job that helps make effective the Nuclear Test Ban Treaty. Other satellites give early warning of missile launchings.

Another type of military satellite, originally used to guide submarines, has been available to navigators of civilian ships since 1967. Six Transit satellites orbit about 700 miles high on a north-south circuit. Every two minutes, they beam coded radio signals to shipboard receivers; computers then translate the information into readings of latitude and longitude.

In Houston during the spring of 1971 I talked with Academician Boris N. Petrov, chairman of the Intercosmos Council, who explained the wide

N.G.S. PHOTOGRAPHER OTIS IMBODEN

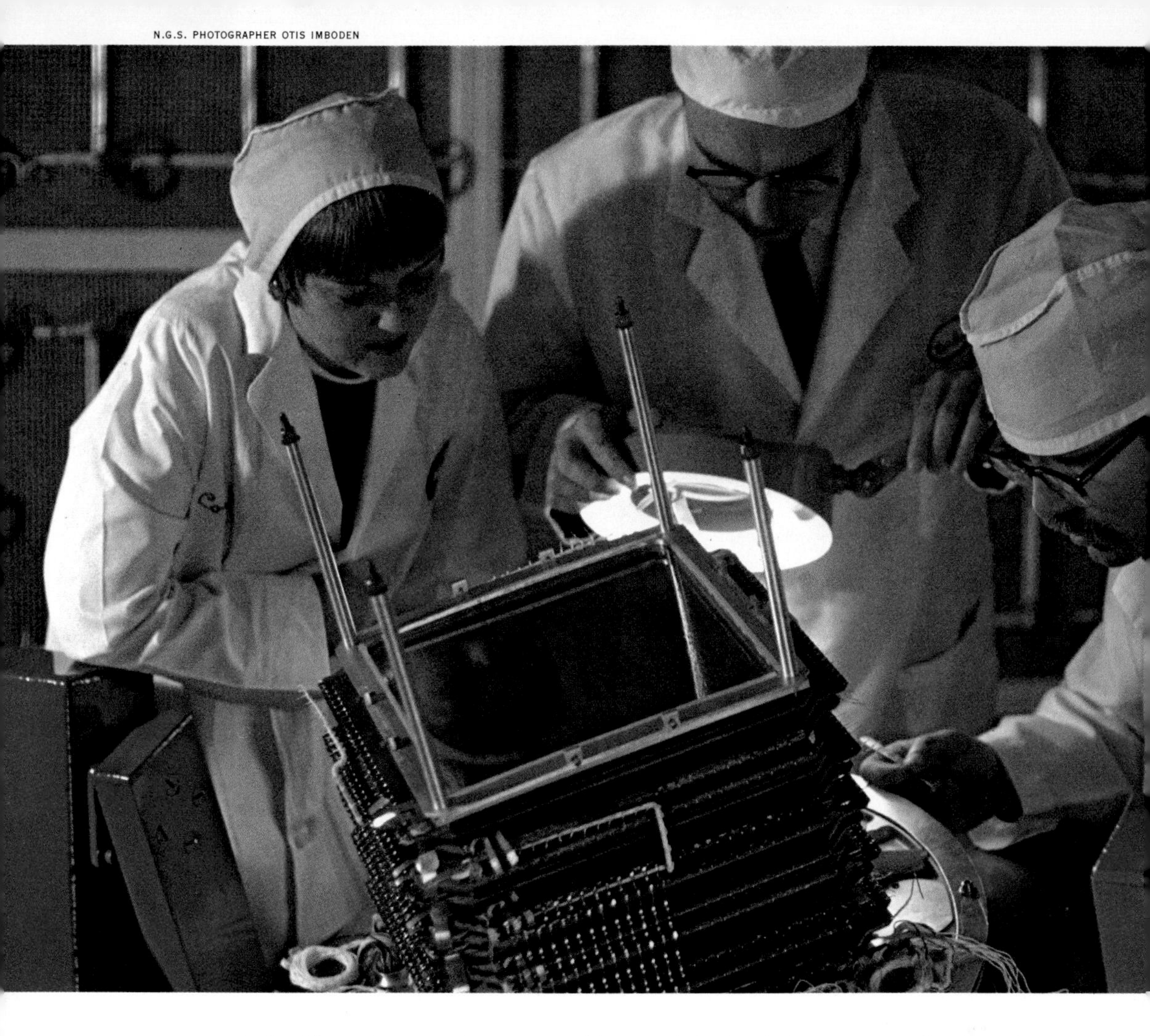

variety of scientific and communications satellites operated cooperatively by the Soviet Union with Czechoslovakia, Poland, East Germany, Rumania, Bulgaria, and Hungary. A leading Soviet exponent of automated spacecraft, Petrov cited among notable achievements the docking in orbit of the unmanned Cosmos 186 and Cosmos 188 in 1967; the worldwide Meteor system, four meteorological satellites operating around the clock; and the 24-hour-a-day Molniya satcoms which relay long-distance telephone and radio-telephone communications anywhere in the Soviet Union.

While communications satellites have been bringing people of widely separated countries closer together, meteorological satellites, or metsats, have been saving lives and property with early warnings of storms. Cloud pictures reveal hurricanes when they first begin to form, and follow their menacing progress day by day.

Satellite photography can yield information that can be used to help protect our own environment. By 1974 the several hundred investigators of the Earth Resources Technology Satellite—the ERTS project—had compiled a wealth of new evidence to help us understand earth's environment and to help geologists in their search for new energy sources and raw materials.

As with their satcoms, the Russians did not announce a metsat, Cosmos 122, until the mid-1960's. During the four-month life of Cosmos 122, cloud-picture exchange by direct cable began between the World Meteorological Centers in Moscow and Suitland, Maryland. After sending up Cosmos 226 in June 1968, the Russians started daily transmission of pictures—7 to 15 mosaics of large areas of the world, along with visual and verbal interpretations. "Their pictures are excellent," said Robert A. Laudrille, communications manager of the National Environmental Satellite Service.

A selection of cloud pictures also goes every day from Suitland to Moscow, chosen from an environmental satellite operated by the National Oceanic and Atmospheric Administration.

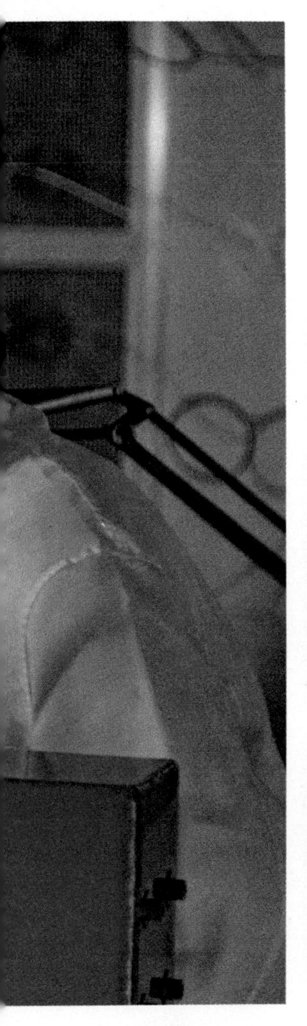

PAINTING BY HELMUT K. WIMMER, AMERICAN MUSEUM OF NATURAL HISTORY, HAYDEN PLANETARIUM

FIRSTS—AND FINALITY: *Mrs. Marjorie R. Townsend, first woman to manage a U. S. satellite-development project—the Small Astronomy Satellite program—inspects a half-built gamma-ray detector with Dr. Carl Fichtel (center) and Dr. Donald Kniffen at Goddard Space Flight Center. Sent aloft in 1972 aboard the second SAS, the instrument measured the intensity and distribution of gamma radiation in space. The first SAS, launched in 1970, detected X-rays emanating from an invisible source near a giant star in the constellation Cygnus. This data, confirmed in 1973 by NASA's Copernicus satellite, points to the existence of a "collapsar" or "black hole." The ultimate concentrations of matter, black holes represent the strangest fate possible for dying stars. Their thermonuclear fuel exhausted, the stars shrink and grow denser until they collapse into themselves. No light escapes, but bursts of X-rays may signal their presence. In the painting above, light spirals into a black hole and vanishes, while a visible red-giant star shines beyond the collapsar's intense gravitational pull. Will the universe end thus? Astrophysicists seek an answer from data on black holes, at once the ghosts and graves of giant stars.*

The satellites store their pictures on magnetic tape; on command, ground stations at Fairbanks, Alaska, and Wallops Island, Virginia, acquire them and send them to Suitland by wire. After they are pieced together into large picture mosaics, the photographs show daily cloud distribution over the entire earth. The satellites also measure the temperature and moisture content of different levels of earth's atmosphere.

Night as well as day cloud pictures came from Nimbus 2 during 1966. Its butterflylike wings glittering with blue-gray solar cells, Nimbus swept in a polar orbit around the earth about 14 times every 24 hours. Nimbus 3, with a supplementary nuclear power supply, was launched in 1969, and photographed the world slice by slice.

The Nimbus 5 metsat, launched in 1972, took pictures of the Antarctic continent directly through the clouds that hang over the polar areas during most of the year. Nimbus 5 also established that during the 1972-73 season there was far more ice in the East Greenland Sea than had been

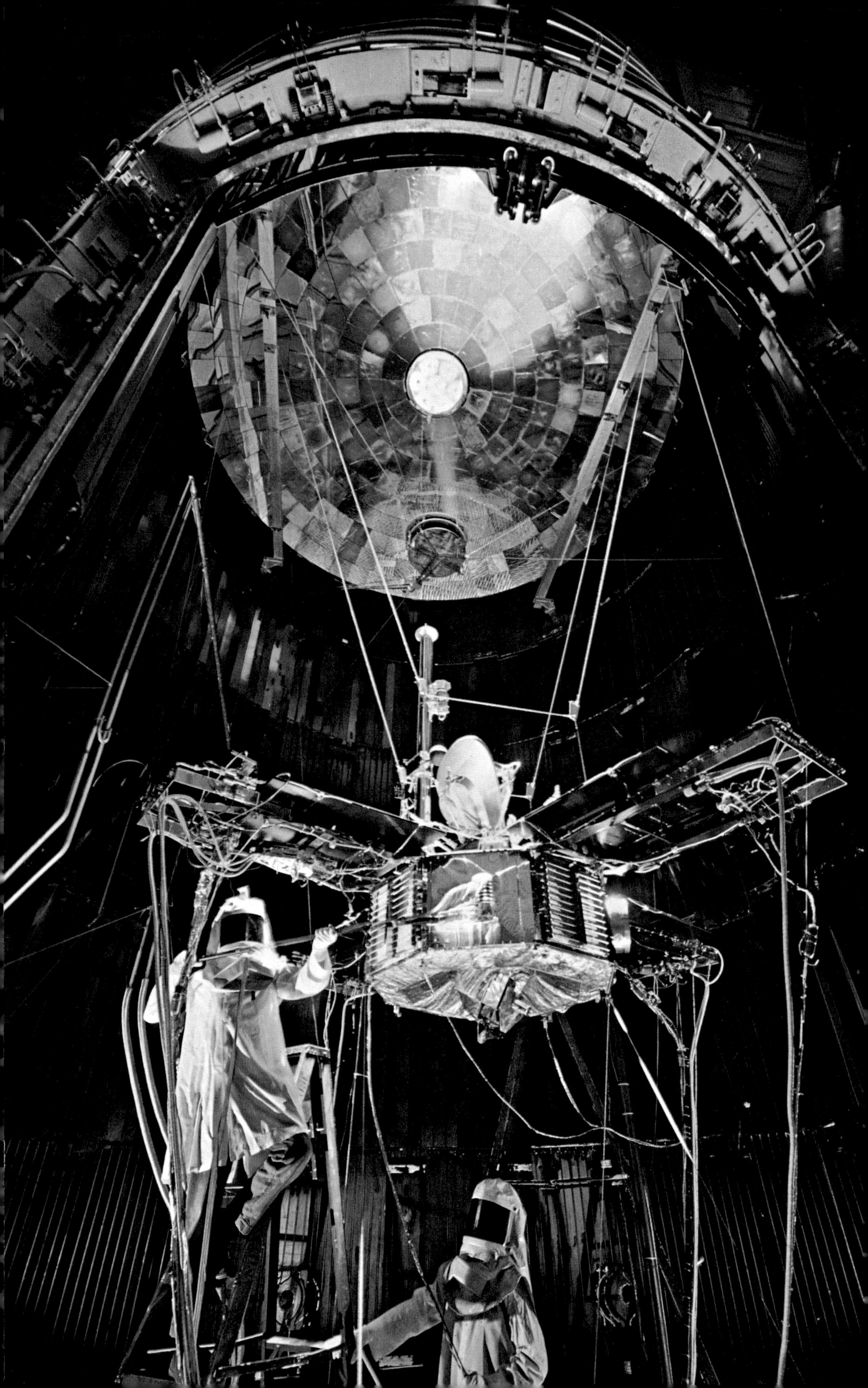

DISTINGUISHED MEMBER *of a successful family—Mariner 4 undergoes preparation for temperature tests inside a space simulator at the Jet Propulsion Laboratory in Pasadena, California. The robot craft transmitted the first close-up views of Mars in 1965. Mariner 9 startled scientists in 1972 with views of Nix Olympica, a 15-mile-high Martian volcano. In 1974, on the first dual planetary mission, Mariner 10 photographed Venus and Mercury.*

estimated. In addition, Nimbus 5 furnished atmosphere and rainfall data worldwide and, for the first time, at microwave frequencies.

Color pictures, helpful in identifying both earth and cloud features, came for a time from ATS-3 and other Applications Technology Satellites. In synchronous orbit, ATS-1 over the Pacific and ATS-3 over the Atlantic continue to test weather, communications, and photographic and navigational equipment. ATS cloud pictures taken every 20 minutes show the earth as a disk, and technicians project the pictures rapidly in sequence to study cloud movement.

While satellites that resemble caldrons, cylinders, and coffee grinders with wings are handling their workaday chores, other craft with purely scientific assignments map space or investigate objects far out in the universe. Their missions may concern astronomy, geodetics, or other fields, and they radio streams of facts exciting to researchers and essential to planners of manned flight.

Explorer 51, launched late in 1973, was NASA's first highly maneuverable unmanned satellite. On command, it could dip as low as 72 miles above the ground, less than five miles above the record altitude reached by the X-15 rocket plane.

Explorer and Pioneer have been familiar names since the first U. S. satellite launchings. Several newer families—with names like OSO, OGO, OAO, GEOS—have joined the others in scientific exploration of the cosmos.

I studied a replica of Pioneer 4 at the Smithsonian Institution and asked Louis R. Purnell, the spacecraft curator, how he would describe it. "Oh, it's sort of a little merry-go-round with a foot-high pointed canopy striped with gold," he said with a laugh. "Rather pretty."

Pioneer 4's rocket booster had also looked rather pretty when I watched it climb above Cape Canaveral in 1959. The probe missed the moon but became the first U. S. orbiter of the sun. Now its remarkably successful grandchildren report on fact-gathering missions around the sun. Since 1965 they, along with ten IMP's (Interplanetary Monitoring Platforms) of the Explorer family and five Orbiting Geophysical Observatories (OGO), have examined cosmic rays, magnetic fields, the aurora and airglow, micrometeoroids, radio noises from space, and dozens of other phenomena. One result of the work of our robots is that charts of the Van Allen radiation region are now far more accurate than were 19th-century charts of the Gulf Stream.

The final IMP satellite was launched in 1973 to help observe the earth's radiation environment through an 11-year solar cycle.

With their scanning spectrometers, Orbiting Solar Observatory satellites have looked straight at the sun since 1962. OSO 7 in 1971 established that the sun's corona at its poles is about a million degrees cooler than the rest of the sun. Also, some mysterious energy-conversion process apparently going on in solar flares creates temperatures that are millions of degrees hotter than either the interior or the surface of the sun, according to Dr. John F. Clark, Director of Goddard Space Flight Center.

The 50-member Explorer family does not confine its talents to physics. In geodetics, Explorer 29 in 1965 began a mission continued by GEOS 2—Geodetic Earth Orbiting Satellite—to examine gravity and the earth's atmosphere.

In astronomy, an Explorer with an X-shaped antenna 1,500 feet from tip to tip began in mid-1968 to study mysterious radio signals from space.

Uhuru, the first of two all-X-ray satellites, was launched from the Italian San Marco platform in the Indian Ocean in December 1970. Within a year it had discovered 35 additional sources of X-ray emissions in our own galaxy alone.

"The most complex of the automated observatories," says Dr. Naugle, "is the OAO—Orbiting Astronomical Observatory. No technical challenge in the space program is more difficult." The eight-sided canister with solar-cell paddles carries several telescopes. Heaviest of all U. S. unmanned scientific satellites, it weighs 4,400 pounds. The first one failed at launch in 1966. In 1968, however, a successful OAO began returning impressive data on stellar phenomena and the nature of our galaxy from an orbit 400 miles up.

With the Orbiting Astronomical Observatory,

J.R. EYERMAN

the range of responsibility for automated spacecraft underwent a sort of quantum jump from investigation of the moon and the planets to include investigation of intergalactic and interstellar space. As the returns from robot investigation of other parts of the universe begin to match secrets unfolding almost daily concerning our solar system, man's new vantage point in space is providing a dramatic extension of our knowledge—both forward in space and backward in time.

An OAO active in 1973, the Copernicus satellite, looked at a strange "black hole" phenomenon about 4,600 light years from earth. According to Dr. Peter Sanford of University College, London, the atmosphere of a visible star was being sucked into the core of an invisible "black hole" of enormous gravity and density. To Dr. Sanford, the doughnut-shaped invisible "hole" was squeezing the stellar atmosphere so hard that it was giving off X-rays.

Information on "black holes" is important, says Dr. Clark because they are "the most immense, irresistible forces in the universe and like gigantic vacuum cleaners, they suck and compress everything that comes near them to a size we can hardly imagine. For example, Manhattan Island—buildings, automobiles, everything—would be condensed to about the size of a pea. But it would still weigh 100 billion tons!"

In their performance the artful robots, much tougher than man, are otherwise daily becoming more manlike. They watch our weather. They guide our ships at sea. They record the characteristics of our land. Like latter-day heliographs they help us send our messages. They orbit and monitor the moon and the sun. They reconnoiter the planets. They peer outward beyond our solar system toward other stars, other galaxies.

Perhaps someday they will be the first to tell us that the flower of earth is not the only flower of life in the vast fields of space.

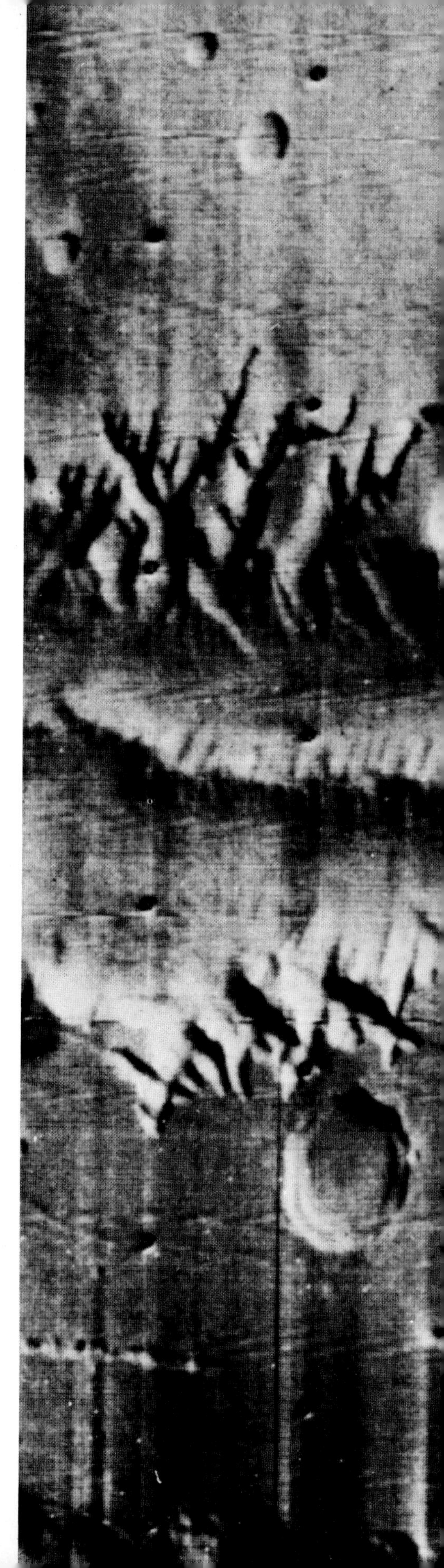

IMMENSE MARTIAN CANYON, *photographed in 1972 by Mariner 9, cuts the red planet's surface 300 miles south of its equator. This portion of the 2,500-mile-long rift system measures nearly four times as deep, more than five times as wide, as the Grand Canyon. The rift, perhaps the result of faulting, indicates that Mars has a geologically active interior. Scientists now consider Mars as being more like the earth than like the moon.*

NASA, MARINER 9

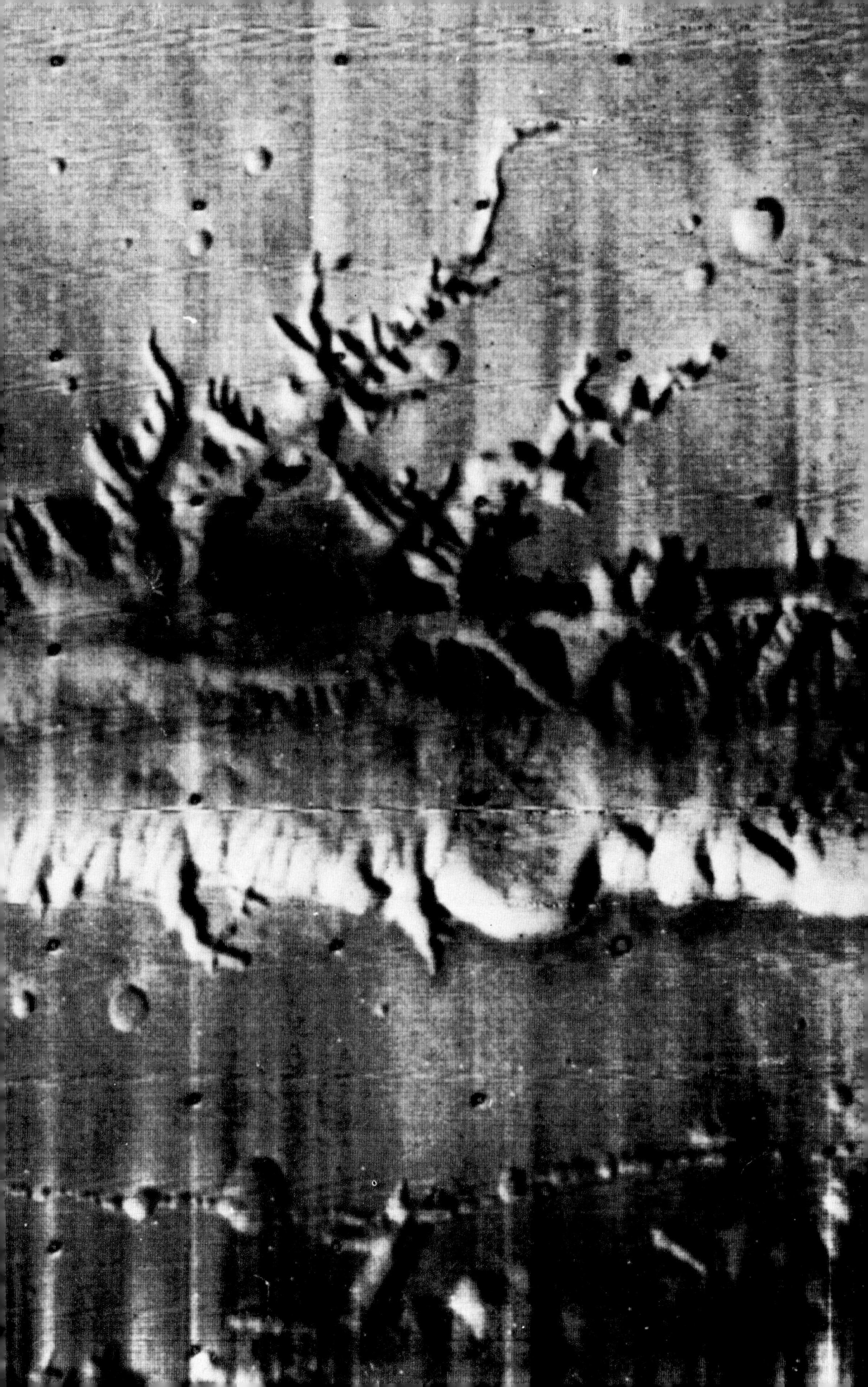

6/ CAMERAS OF THE COSMOS THE VIEW FROM SPACE

On the final Skylab flight, scientist-astronaut Dr. Edward G. Gibson suddenly leaned forward against his seat restraint. For days he had been watching the sun through the Apollo Telescope Mount (ATM). Now he saw a slight turbulence along the western side of the sun. Focusing ATM's cameras and other instruments, he made a unique recording of the "flash phase" of the beginning of a solar flare. Dr. Gibson's alertness at his Skylab post was but another of many examples of the skill and dedication of the photographer in space.

Beginning in May 1973, the three Skylab missions involved the most extensive use of sophisticated photographic and remote-sensing equipment ever employed in space. These three manned missions, totaling 171 days in earth orbit, furnished 46,146 still pictures of earth, 175,047 still pictures of the sun, the stars, and the comet Kohoutek, plus thousands of feet of motion-picture footage and more than 45 miles of data-laden magnetic tapes.

The earth pictures alone would keep analysts occupied for years. Skylab revealed some startling views. Among them were strange counterclockwise cloud vortices over the Mexican island of Guadalupe, a revealing view of Hurricane Ellen that showed the domed tops of cloud formations, the nearly perfect symmetry of a New Zealand volcano, midwinter surface ice patterns in the Gulf of St. Lawrence in Canada, the first comprehensive views of drought-stricken areas in Africa and Australia, previously unknown ocean cold-water eddies around major land juts and capes, and so-called plankton blooms in the Falkland Current off Argentina. The third Skylab crew was aloft long enough to see and photograph the color change in the vast wheat-growing areas of Australia and Argentina.

Some of the most useful images to come from Skylab were produced by cameras that simultaneously registered full color, infrared, and four narrow bands of the spectrum, including infrared

INDIAN SUBCONTINENT *sprawls 622 miles below Gemini 11, September 14, 1966. An 8-foot antenna protrudes in the foreground. Thousands of such photographs, taken during manned and unmanned missions, give oceanographers, weathermen, and geographers valuable new views of our planet.*

RICHARD F. GORDON, JR., NASA

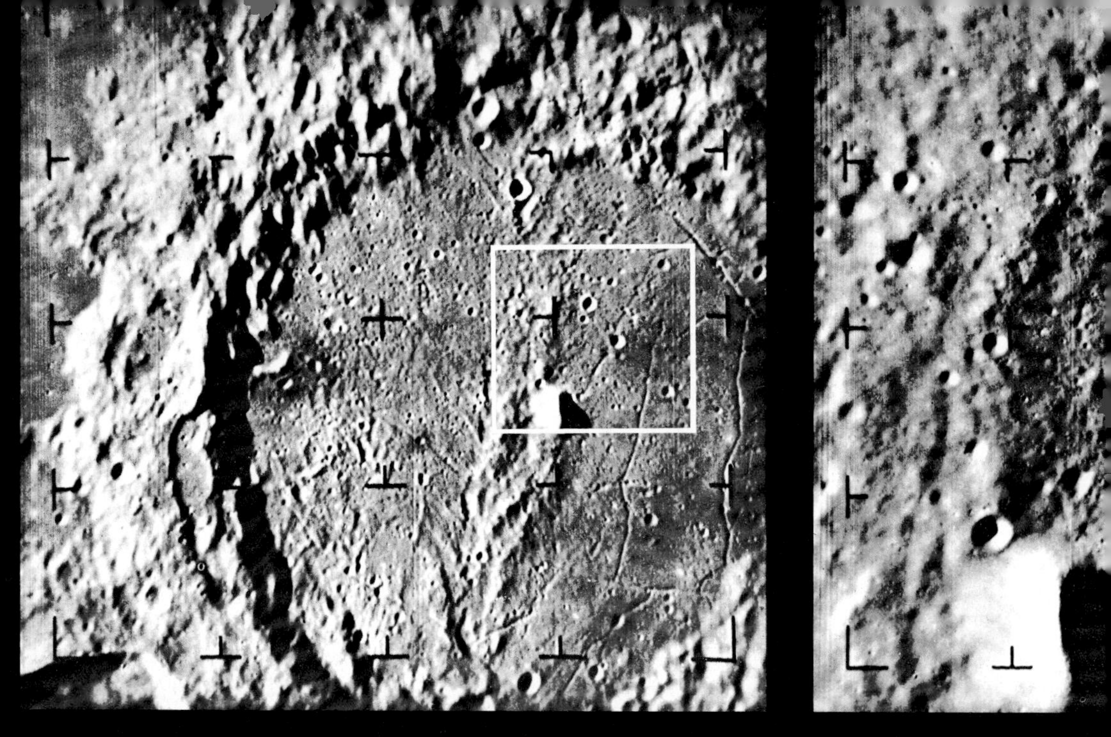

Huge crater Alphonsus looms before the cameras of Ranger 9, hurtling toward the moon at 6,000 miles an hour. Launched from Cape Kennedy the afternoon of March 21, 1965, the craft began transmitting pictures some 65 hours later; within 20 minutes—before impact just three miles from its intended target—Ranger 9 sent 5,814 pictures "live from the moon" to television screens on earth. The central peak of the 70-mile-wide crater rises 3,300 feet.

and visible light. By superimposing the images, or comparing them side by side, scientists obtained a diversity of information. For example, they were able to measure soil moisture and snowfall, to determine whether crops were diseased or healthy, to look for signs of minerals or oil, and to check for pollution, flooding, and erosion.

Although the comet Kohoutek was a visual disappointment on earth, to the third Skylab astronauts it was spectacular. "The sky was so black and the comet was so brilliant," said Gerald P. Carr, "it was . . . a gorgeous sight."

Gibson photographed the "spike" of Kohoutek, a streak of illuminated dust shooting out in a different direction from the familiar tail. Views of the spike led to speculation that not all of Kohoutek's tail was whipped around behind it as it looped close to the sun; some of the tail may have been left as a luminous residue.

Both from earth orbit and from distant space, cameras have recorded vistas never before seen, unhampered by atmospheric distortions. In addition, cameras in lunar orbit and manned cameras on the lunar surface have given us thousands of vivid images of our nearest neighbor in space.

A few days after Apollo 15 explorers Dave Scott and Jim Irwin returned from the moon's Apennine Mountains and Hadley Rille, geologists pored over their photographic treasure. Dr. William Phinney, chief of the Geology Branch at the space center in Houston, called the photographic results a major bonus.

"We got as much information this time from the photographs as from the rocks," he said.

Later, on Apollo 17, Astronaut Gene Cernan determinedly went after a single picture: To include in one frame fellow crew member Jack Schmitt, a huge boulder, and the lunar Rover, all against the dramatic background of the Taurus-Littrow valley, Cernan laboriously climbed an incline and aimed and re-aimed until he was satisfied that he had the right perspective. His patience and expertise helped return the most professional color photographs ever taken on earth's natural satellite.

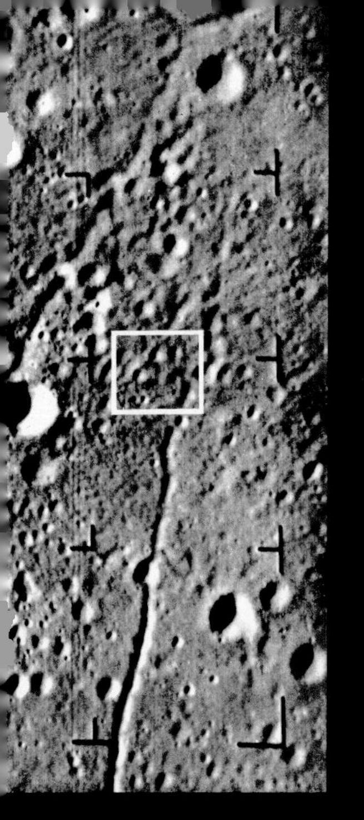

NATIONAL AERONAUTICS AND SPACE ADMINISTRATION

White lines on the photograph at left, taken from 163 miles, mark the area of the center picture, taken at 42 miles. The crater at the upper right of the square measures less than half a mile across. Long, deep depressions called rilles furrow the surface. The small square encloses the area of the last complete image (right), transmitted from 4.47 miles seconds before impact at the spot marked by the arrow. Craters 40 feet wide appear pinhole size.

An initial and compelling photographic objective in space was to determine the nature of the lunar surface to guide scientists and engineers in the design of the moon-bound Apollo spacecraft. Without such knowledge, the specifications of the lunar module might be incorrect. The development of the four saucer-shaped footpads, for example, might be based on a faulty estimate of the bearing strength of the moon's surface.

To find the answers, the United States devised, first of all, the Ranger program for hard-landing on the moon. The Jet Propulsion Laboratory spent the better part of five years and more than 200 million dollars on the series of ambitious flights.

Prior to the Ranger shots, according to astrogeologist Eugene M. Shoemaker, "Our best telescopes had brought us visually to within 400 miles of the moon's surface. But they could do no more. . . . No lunar detail less than 800 feet across had been distinguished from our planet."

The plan for Ranger 7 was a bold one. The 800-pound probe would be sent on a collision course toward a plainlike region 390 miles south of the great crater Copernicus. Beginning just minutes before the spacecraft's impact, six television cameras would transmit the historic view until they were dashed to pieces on the moon.

On July 31, 1964, sleepless personnel in space centers across the country waited expectantly. At the Jet Propulsion Laboratory secretaries came to work two hours early to witness the climax of the 67-hour flight.

A hot line from the Goldstone tracking station in California's Mojave Desert to JPL passed the first word that the pictures were coming in. A voice called out: ". . . 10 seconds . . . we're receiving pictures to the end . . . impact . . . impact!" The sudden silence was followed by shouts, handshakes, and wild cheering. America's photographic mission to the moon was a success.

The 4,300 images transmitted in the final 17 minutes were the first finely detailed pictures of the moon's surface. Run on film, like a movie, they gave a remarkably realistic impression of an impending crash. The film was later shown to the astronauts in Houston. Just when impact seemed

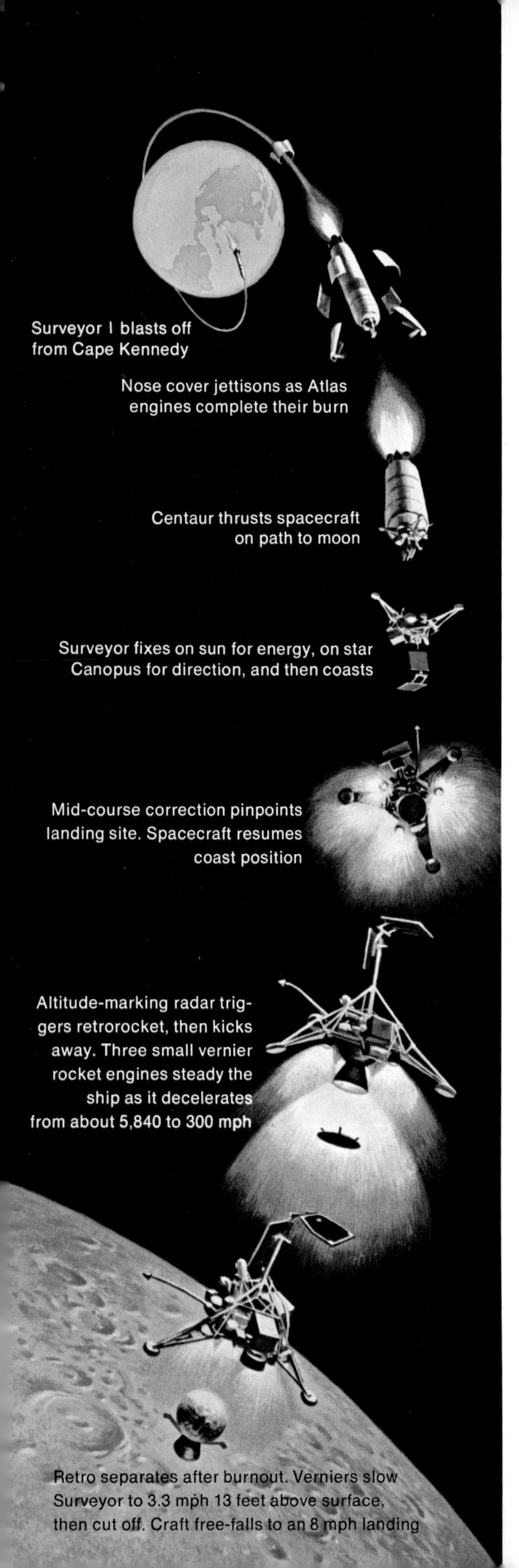

ILLUSTRATION BY DAVIS MELTZER (LEFT); NASA (BELOW); ALAN L. BEAN, NASA

ENCOUNTERING A SPACE "RELIC," *Apollo 12 Astronaut Charles Conrad, Jr., inspects the television camera on Surveyor 3. The unmanned craft landed on the moon 31 months before* Intrepid *(on lunar horizon) arrived in November 1969. Conrad and Alan L. Bean set their spacecraft down 600 feet from Surveyor 3; they found it dusty but intact, and snipped off parts for study on earth. The first Surveyor (sequence at left) landed on the moon June 2, 1966. After a three-day journey from Cape Kennedy, the mechanical "cameraman" transmitted the photograph above of a shadow-filled crater about 10 feet in diameter. The seven Surveyor flights demonstrated that astronauts would find firm footing when they landed on the moon.*

imminent, Wally Schirra broke up the house by shouting, "Bail out, you fool!"

The next epochal stage in lunar photography began in 1966 with soft-landings on the moon by the United States and the U.S.S.R. of packages containing cameras and other instruments. Earlier, in 1959, Soviet scientists had succeeded in photographing the moon's far side with their Luna 3 probe, but the pictures had limited value because of poor resolution.

A soft lunar landing was not an easy undertaking. The problem facing scientists of both countries was formidable: Fire the braking rockets of an incoming lunar probe too soon and the spacecraft would stall too high and fall too hard for its instruments to survive; slow it down too late and it would shatter in a high-velocity crash.

EUGENE A. CERNAN, NASA

On February 3, 1966, the Russians finally succeeded in this delicate braking maneuver. Luna 9, a spherical automated station weighing about 220 pounds, came safely to rest in the Ocean of Storms. In the weak gravity of the moon, the capsule and its instruments weighed just over 36 pounds. A radio command opened the lens of Luna 9's television camera, and the resulting panel of nine pictures initiated the era of on-the-surface scientific investigation of the moon.

Unknown to the Russians, British scientists had borrowed a telephotographic facsimile machine from a newspaper and connected it with their radio telescope at Jodrell Bank. They then tuned to the frequency announced by the Russians. When this hasty improvisation worked, the British actually released the pictures to the world before the astonished Russians did.

Then an extraordinary thing happened. When pictures started coming in on February 5, it was obvious from a shift in the horizon angle that Luna 9 had moved or lurched during the earth night. Was this evidence of a lunar quake? Was the moon still volcanic? Or did meteoroid impacts cause the camera to change position slightly? No one knew for sure. Later I asked Academician Alexander Lebedinsky of the Moscow Planetarium what had happened. He could only speculate that

HUGE BOULDER *dwarfs the lunar Rover and Apollo 17 Astronaut Harrison H. Schmitt, who retrieves a device that indicates color and scale for photographs in the Taurus-Littrow valley. The rock once perched high on a mountain. Probably dislodged by a meteorite, it tumbled down and split into five segments as it came to rest. "A geologist's paradise, if I ever saw one," Schmitt called the valley.*

the lurch was due to a breakdown of the soil under the weight of the capsule.

Three and a half months after Luna 9 soft-landed, the U. S. Surveyor 1 gently settled its 600 earth-pounds in the vast Ocean of Storms. By now, both countries had also succeeded in placing cameras in moon orbit—the United States with its Lunar Orbiters (each containing a 150-pound radio-controlled camera), and the Soviet Union with a continuation of its orbiting series that began in April 1966 with Luna 10.

By early 1968, clear and usable Ranger, Surveyor, and Lunar Orbiter photographs numbered more than 100,000. Additionally, the Surveyor spacecraft had carried out moon-surface exploration of five areas previously chosen as possible landing sites for the manned Apollo Lunar Module.

"For 50 years," Dr. Shoemaker told me, "man has been speculating about the nature of the lunar

(Continued on page 132)

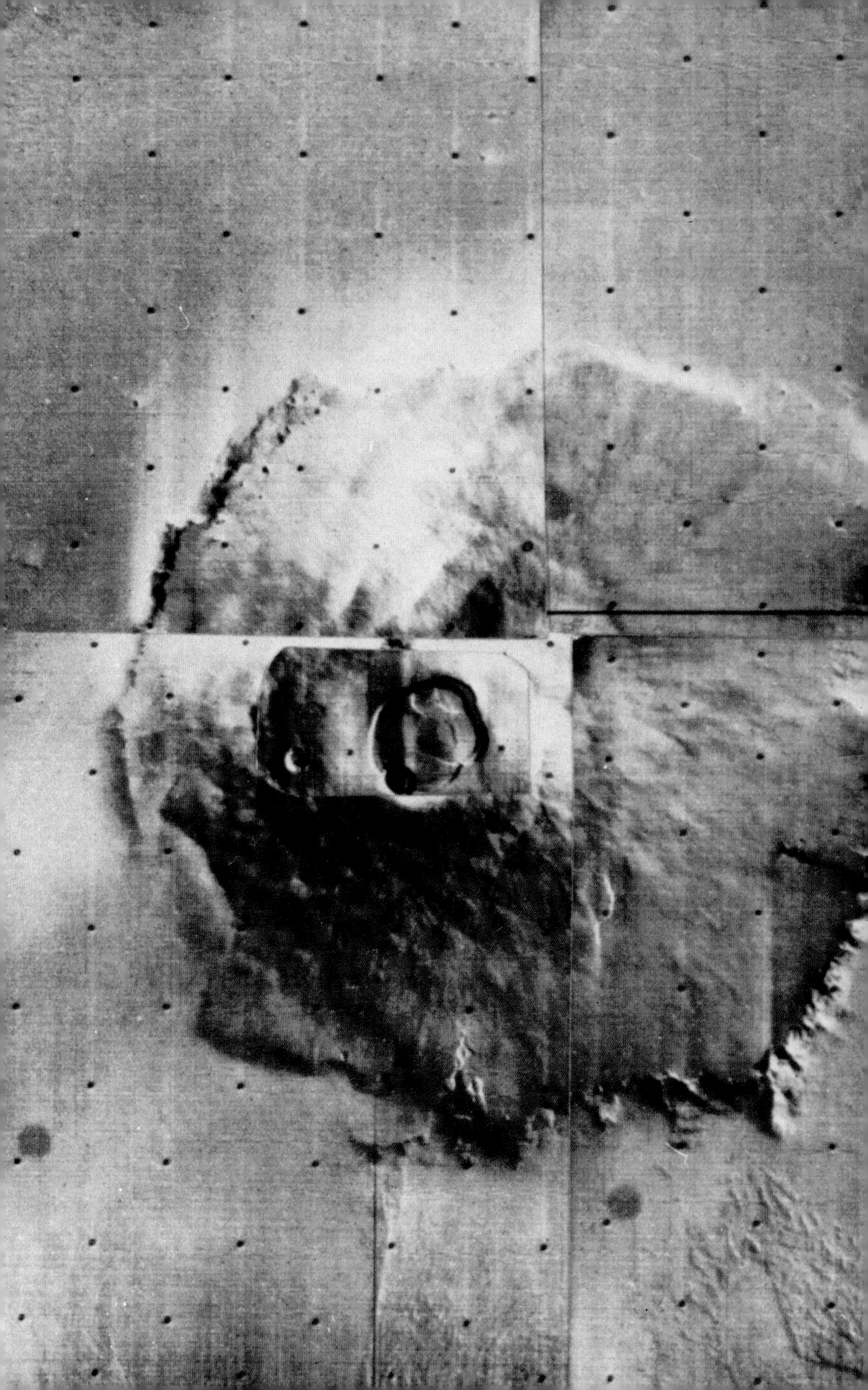

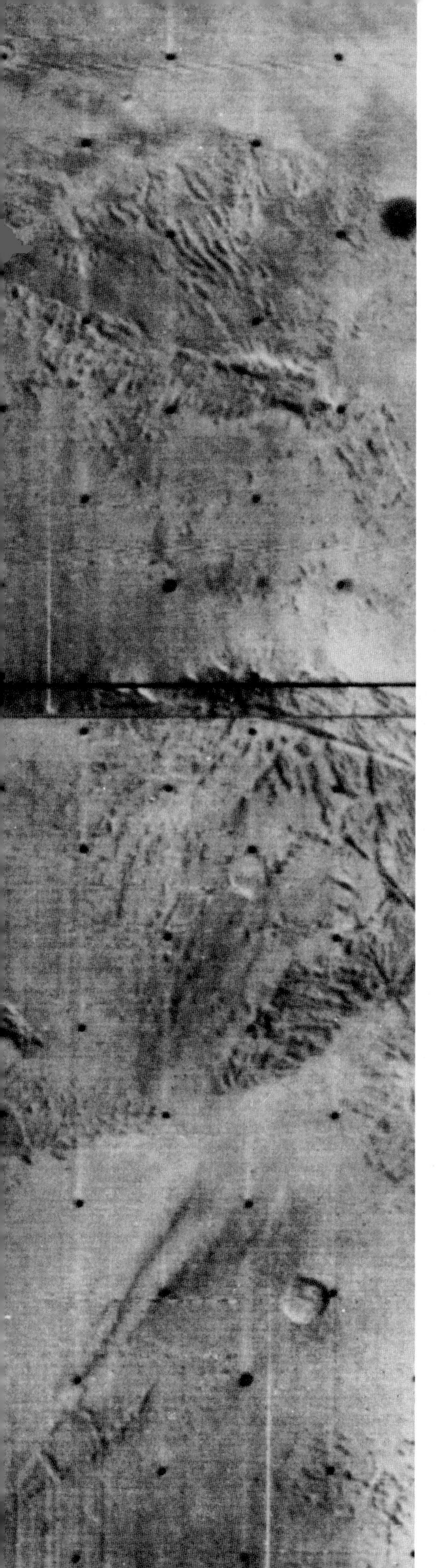

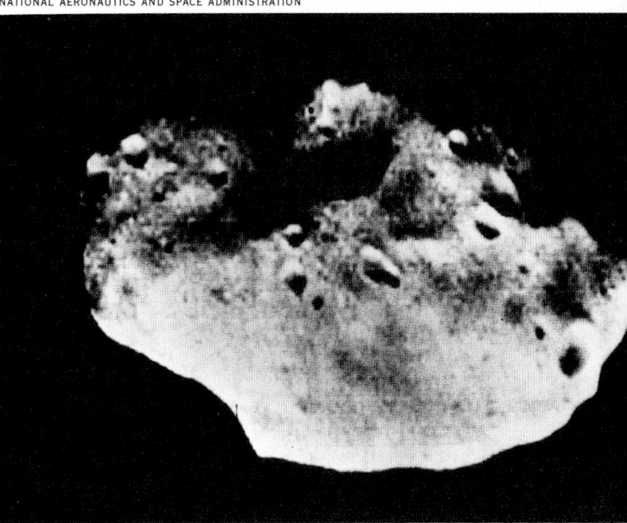

MARTIAN MOON *Phobos shows a cratered face in this photograph transmitted in December 1971. Mariner 9 captured the image from orbit, 3,444 miles away. Twice the size of Deimos, Mars' other moon, Phobos measures 16 miles across.*

MARTIAN VOLCANO *Nix Olympica rises from a plain more than 1,500 miles below Mariner 9 in this mosaic of photographs taken in January 1972. The main crater, at the summit, reaches 40 miles from rim to rim. The base measures 310 miles across, covering an area larger than New England.*

ROBOT EXPLORER *Mariner 9 reaches the goal of its 167-day journey from earth on November 13, 1971. In four and a half months, the craft's two TV cameras took nearly 7,000 pictures of Mars.*

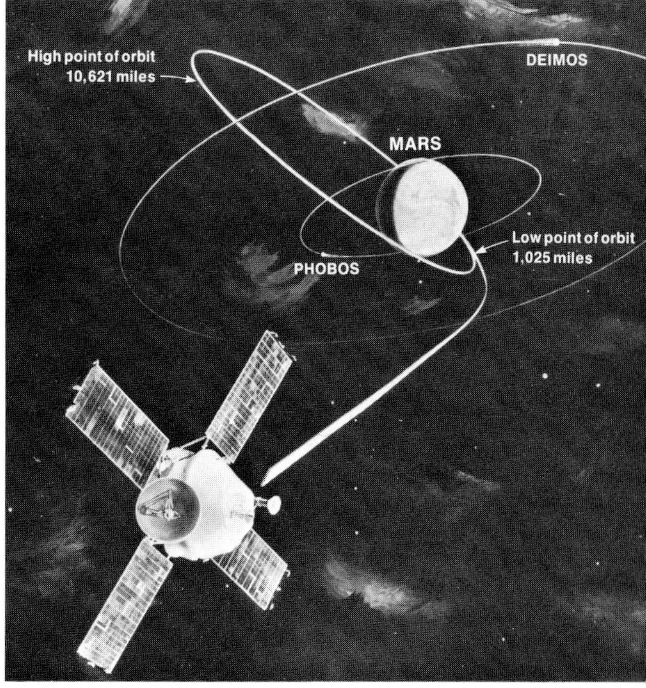

JAMES A. LOVELL, JR. (BELOW); ALL NASA

SPACE-ROVING EYES PROVIDE A LOFTY VIEW OF EARTH

Horizon curves below Astronaut David R. Scott as he stands at the hatch of the Apollo 9 command module 150 miles up. "Boy, oh boy, what a view!" exclaimed Russell L. Schweickart, pilot of the lunar module, as he took the photograph. Although they and James A. McDivitt devoted much of their ten-day flight in March 1969 to photography, their main mission consisted in maneuvering and docking the lunar module in preparation for the Apollo 11 flight to the moon. At center above, Houston sprawls around Galveston Bay in detail so sharp it reveals the Manned Spacecraft Center, individual buildings, even streets. Snowy peaks of Afghanistan's remote Hindu Kush (above, right) rise over dark valleys that provide watercourses for spring runoff. Some three years earlier while orbiting in Gemini 7, Astronaut Frank Borman aimed his camera toward Algeria's Tifernine dunes (right) 150 miles below. These thousand-foot-high waves of sand show one of the ways winds have shaped the surface of the earth. In November 1966, Gemini 12 coasted high above Egypt, the fertile Nile valley, and the Red Sea (far right). Cirrus clouds, caught in the rush of the jet stream at altitudes of 20,000 to 40,000 feet, follow the path of these invisible winds as they circle the globe from west to east at speeds as great as 300 mph.

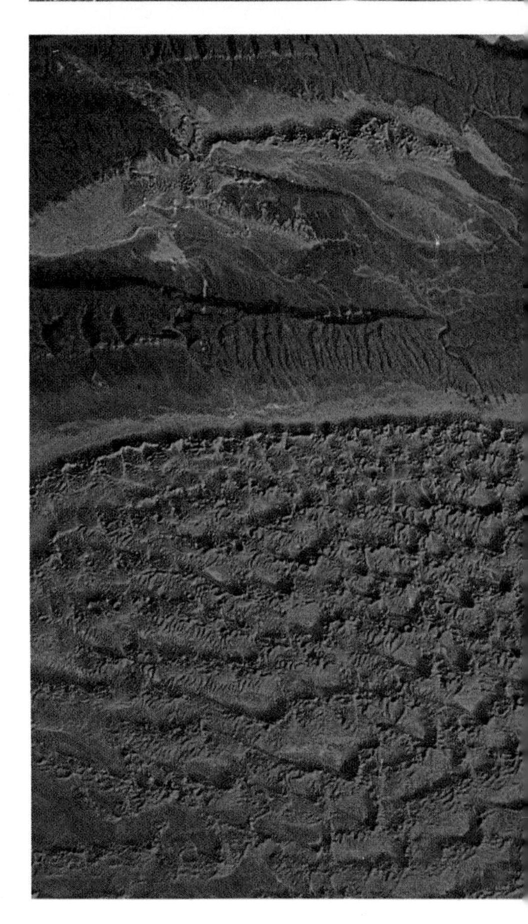

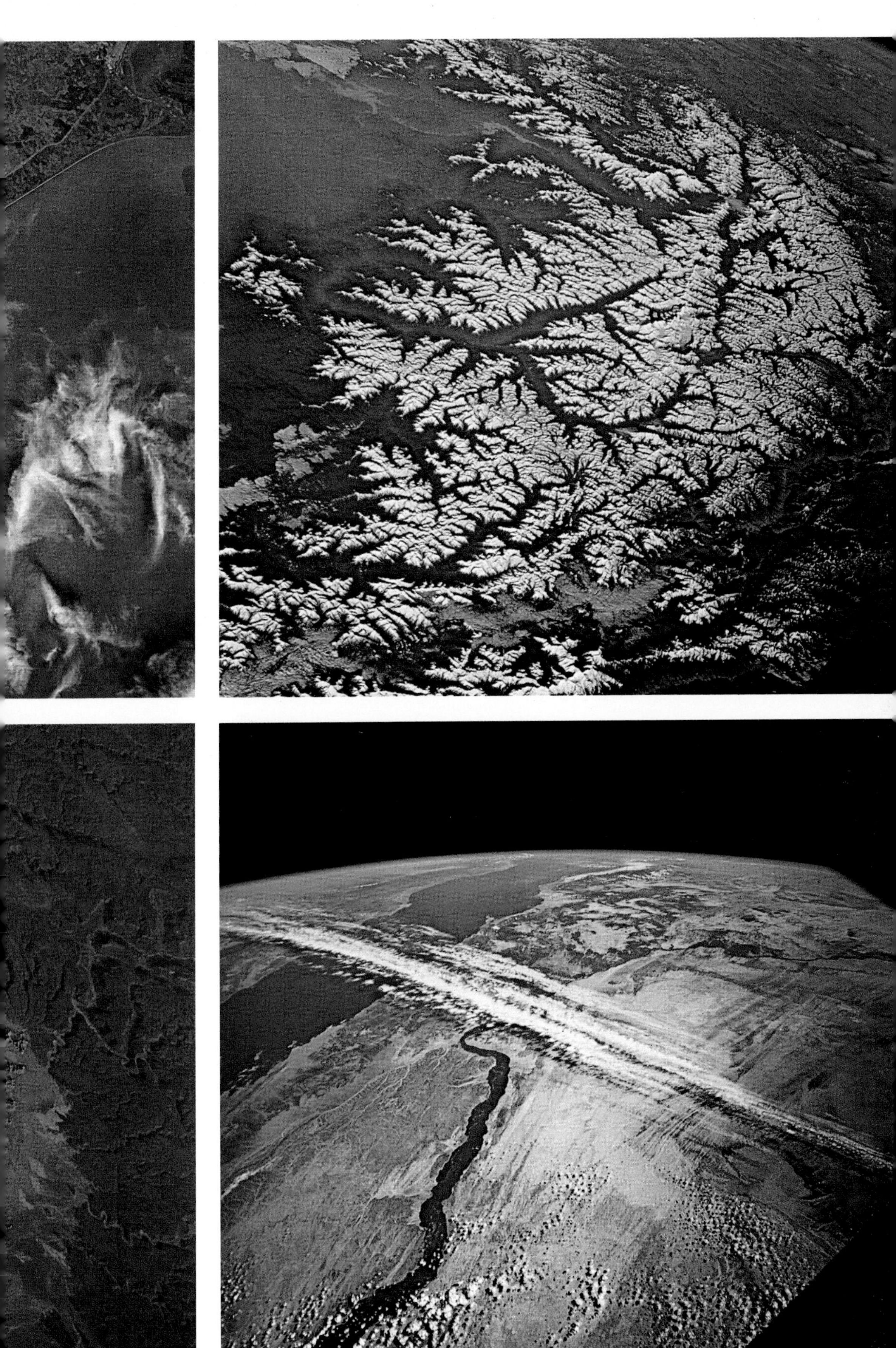

surface. Now, thanks primarily to these photographs, we know that at least 99 percent of the surface is composed of fragmented debris. We know its approximate thickness, its bearing strength, its grain-sized texture, and its cohesion. Surveyor told us the basic chemical composition of the lunar plain is basaltic like that of the most common lavas of earth. Scientists are now in 90 percent agreement that the surface fragments are the cumulative product of prolonged meteoroid bombardment of the moon."

Since the beginning of the Space Age, photography has become one of the scientist's most important means of gathering and preserving information. The Manned Spacecraft Center provided astronauts with training in terrain and weather photography and gave them extensive practice in using the Hasselblad 500C, the Maurer 77-mm Space Camera, and other special equipment. So demanding were photographic duties in space that most astronauts heartily sympathized with Gordon Cooper, who made the first comprehensive study of the earth's "night glow" during his flight in *Faith 7*. After clicking the shutter "all night long," Cooper drawled from space: ". . . all I do is take pictures, pictures, pictures."

In 1961, the Russians provided the first televised view of a cosmonaut as he actually appeared in orbit. Later, the magic of miniaturized television components enabled almost the entire world to witness space walks, rendezvous and docking, the moon as seen from lunar orbit, and Neil Armstrong's first step onto the moon's surface. And with Apollo 15 came color television pictures from a camera mounted on the first automobile on the moon. Each time the Lunar Rover stopped, Edward I. Fendell, a NASA engineer working at his 15-button console a quarter of a million miles away, manipulated the TV camera through traverses, pans, zooms, and wide-angle panoramas. Fendell and his remote camera performed so expertly that engineers and geologists in Mission Control could take useful Polaroid snapshots of the pictures appearing on their TV monitors.

The U. S. attention to space photography has produced an almost incredible record of the first period in history in which man could see his own planet from space. The views of the continents, the panoramic sweep of the seas, and, above all, the capture of the full range of weather—from serene to darkly ominous—comprise an unprecedented scientific and esthetic treasury.

And it is a treasury that is already bringing economic benefits. Geologists use the photographs to prospect for oil and other minerals, to study erosion along coastlines, and to detect changes in the courses of rivers. Oceanographers use them not only in studies of marine biology, including fish migration, but also to investigate the way ocean currents erode the floor of the sea.

Turning both manned and unmanned cameras away from earth, scientists have focused on such phenomena as zodiacal light, sunlight reflected from dust particles orbiting the sun; the mysterious *gegenschein*, a ghostly, dim patch of light—also believed to be interplanetary dust reflection but always located exactly opposite the sun; and the solar corona as seen by a 70-mm Hasselblad camera designed to clarify the processes by which energy flows from the life-giving sun. Computer-aided photography also disclosed the moonlike appearance of Mars.

Both from earth orbit and from distant space, cameras have recorded vistas never seen before, unhampered by the volatile atmosphere where particles and swirls distort the view. From space, as Gordon Cooper pointed out, lights on earth twinkle; the stars remain steady and unblinking.

It is this outward view that many scientists feel is ultimately the most valid reason for sending photographic recorders beyond the atmosphere. Much of what man himself will eventually see as he ventures deeper into space, he will first see on film. Whether that clear distant image falls on cameras and spectrographs in earth orbit or on semipermanent instruments implanted to gaze outward from the far side of the moon, it is certain to give man simultaneously both knowledge and the incentive to reach for knowledge. For good pictures show us not only where we have been, but also where we are going.

INFRARED VIEW *of the Mississippi River above Vicksburg, taken from orbit during Apollo 9's terrain-photography experiment, reveals the recent geologic history of the stream. Louisiana lies at left, to the west of the river; popcorn clouds shadow the valley floor to the east in Mississippi. Such photographs help assess flood-control requirements, and reveal traces of the river's course for as far back as a thousand years.*

NATIONAL AERONAUTICS AND SPACE ADMINISTRATION

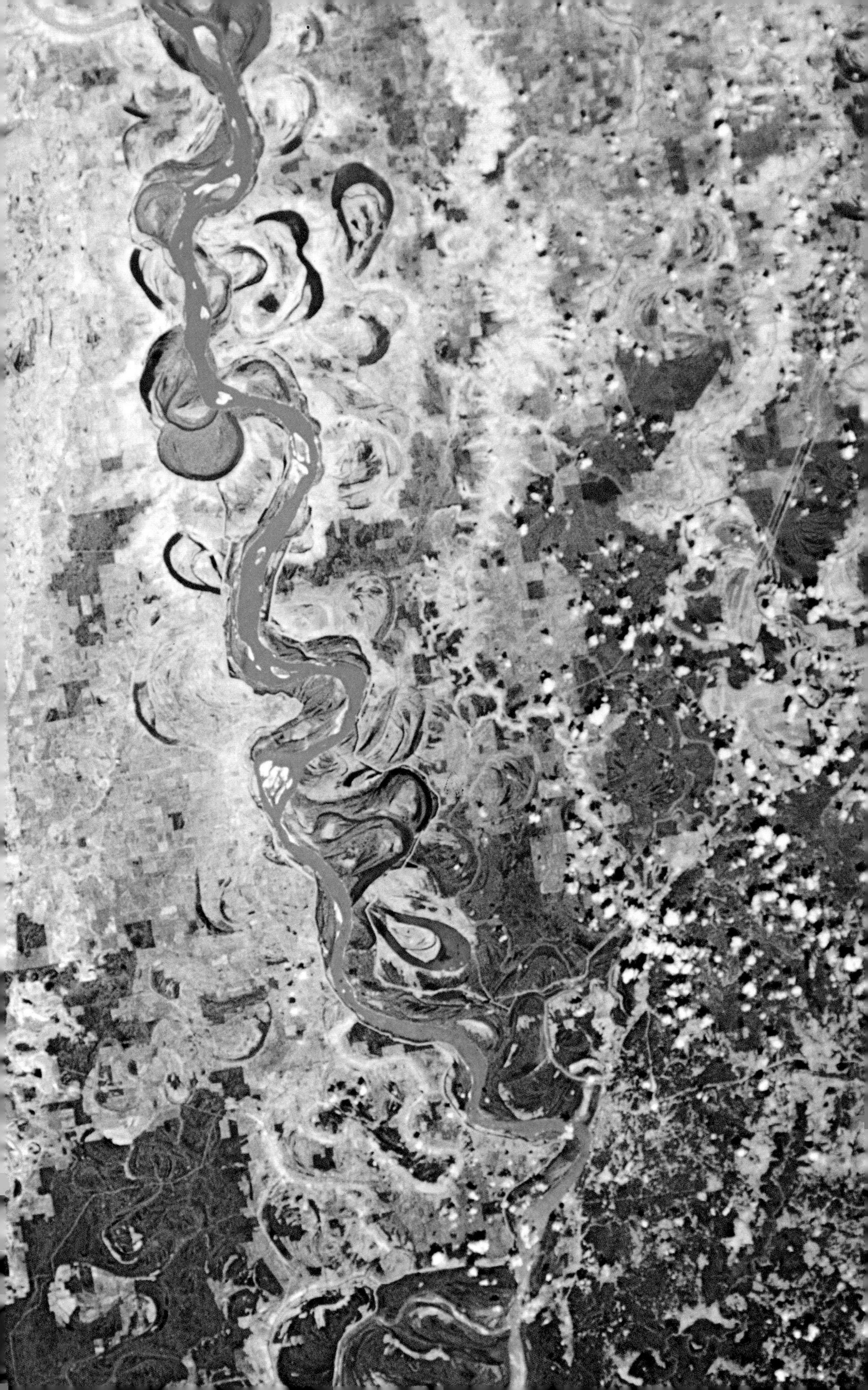

7 / REACHING THE MOON: A DREAM BECOMES REALIT[Y]

The shadowy, blurred figure moved cautiously down the ladder and paused. Five hundred million people peered intently at their television sets, awaiting the unforgettable moment. Then Apollo 11 commander Neil Armstrong placed his left boot firmly on the gray surface, then his right—and a man stood on the moon!

It was, in his concise phrase, "one small step for a man, one giant leap for mankind." It was the climax of decades of effort involving, finally, thousands of men and women. And it was witnessed by one of every seven people on earth.

In the press room of the Manned Spacecraft Center at Houston, my eyes were locked on the mammoth television screen. But my memory was racing back to a day in 1964, even before the Gemini flights, when I had telephoned Neil and Jan Armstrong's home in El Lago, Texas, not far from where I lived. I told Neil that I had prints of the first set of Ranger 7 close-up photographs of the moon. "A friend at the Jet Propulsion Lab just air-expressed them to me. Would you like to see them?"

"Come on over!" was the immediate response.

After I produced the pictures, Neil examined each one as if it had been a bit of the moon itself.

Some scientists believed then that the moon was covered with a thick layer of dust that could be a threat to the proposed landing of Apollo. Neil's initial reaction was one of relief.

"See the fairly sharp edges on these craters?" he pointed out. "If there were a lot of dust over them, I think the craters would be more rounded and less sharply defined. I think when the first man steps out on the moon his footprints will barely sink into the surface." Five years later, after observing that the lunar module footpads "are only depressed in the surface about one or two inches," Neil confirmed with his own footprints the accuracy of his prediction.

My musings were swept away by the activity in the press room. Hundreds of reporters were dashing to their open telephone circuits.

MAN ON THE MOON: *Edwin Aldrin steps from the ladder of Apollo 11 lunar module* Eagle *some 15 minutes after Neil Armstrong's "giant leap for mankind" on July 20, 1969. His pack contains life-support systems that provide oxygen, temperature control, radio, and electrical power.*

NEIL A. ARMSTRONG, NASA

Compelling as their message was, it would still take most of us a long time to comprehend the immensity of the effort that had put men on the moon, and to recognize the obstacles that had been overcome.

Even to begin to turn dream into reality, engineer-designers had to face two crucial challenges: to develop a booster train powerful enough to propel a 50-ton payload to the moon, and to build a spaceship that would function as a home and vehicle for man during the trip there and back.

Secondary challenges were no less difficult: perfecting computer-controlled guidance and navigation accurate enough to land the moon ship; mastering the technique of entering and leaving both earth and lunar orbits, of rendezvousing and docking with another object in space, and of re-entering the near-earth atmosphere; fashioning a space suit to permit a man to work both inside and outside his ship; overcoming the problems of moving and working in the weightlessness of space flight as well as in the weak gravity of the moon; setting up worldwide communications; perfecting spacecraft recovery techniques; finding ways to handle food; and solving the myriad details involved in the operation of a spacecraft.

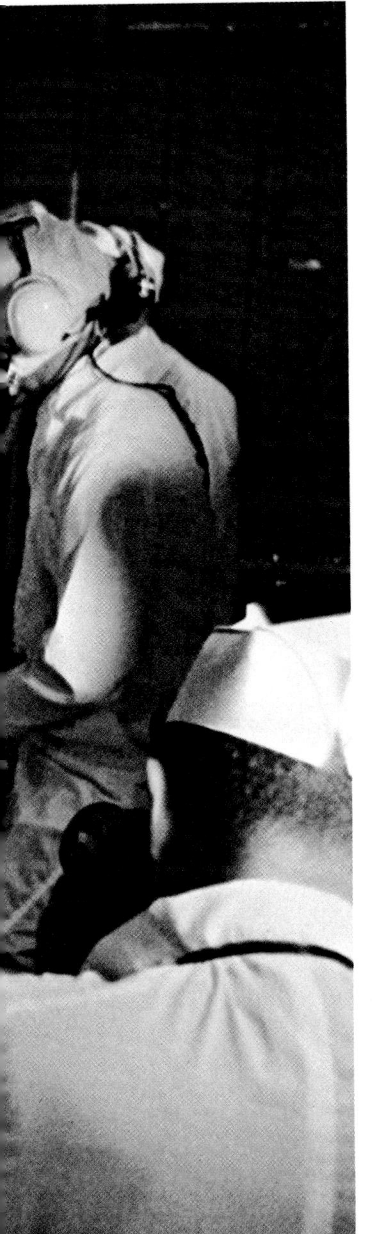

MAJOR MILESTONE *in the reach for the moon: Astronaut John Glenn climbs into* Friendship 7 *during a final check of his one-man Mercury spacecraft before lift-off on America's first earth-girdling mission. Sunset gilds his face in a photograph made by an automatic camera during the three-orbit flight in 1962. Physician William K. Douglas waits through the tense moments of re-entry. A malfunction resulted in the fiery disintegration of the retro-rocket pack, leading Glenn to exclaim, "... that was a real fireball, boy!"*

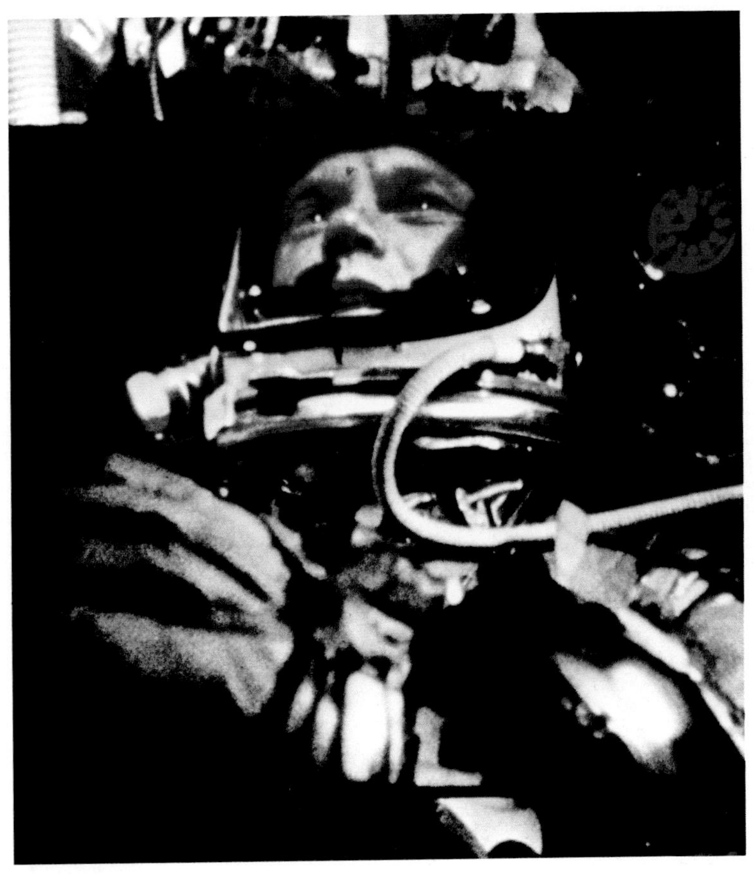

Space engineers had to find new structures and materials to withstand unprecedented shocks, vibrations, and heat. And to power gigantic boosters they had to develop new rocket propellants.

Finally, program administrators of both the U.S.S.R. and the U. S. had to prepare the greatest variable of all—man himself, at once the weakest and strongest link in the mastery of space. Astronauts and cosmonauts attended courses in astrophysics, propulsion, astronomy, geology, and dozens of other subjects. They studied every component and system of the booster and vehicle. They spent hour upon hour in mission simulators, week after week on field trips, in rigorous physical conditioning, and in survival training.

Both countries worked out early major solutions in an amazingly parallel time. Yuri Gagarin opened the first phase of exploration, the one-man flights, on April 12, 1961—a date now commemorated as Cosmonaut Day in the Soviet Union. With the suborbital flight of Alan Shepard in *Freedom 7* just days away, the shy young Russian became the first of the new instant heroes of space. The Vostok 1 flight and Shepard's Mercury mission 23 days later demonstrated that both nations were on the right track in concepts of booster and spacecraft design and guidance and re-entry procedures. The six Vostok and six Mercury flights gradually extended and tested these achievements.

In late 1964, the three men aboard Voskhod 1 used no pressure suits, indicating a progression to more reliable spaceships, and for the first time tested braking rockets for cushioning the landing. With a physician aboard, the flight returned an enormous amount of biomedical data.

The first manned Gemini flight of Gus Grissom and John W. Young established that orbital maneuvers were possible. Then the twin space-walk flights of the United States and the U.S.S.R.—a mere 11 weeks apart—showed that man could venture from his spacecraft.

To judge by the reactions, "walking" or "swimming" in space was an almost intoxicating experience. Cosmonaut Alexei A. Leonov later commented, "The boundless expanses of outer space unfolded before me in their indescribable beauty. . . . I felt fine, was in excellent spirits, and did not want to part with free space."

Ed White, overextending his stay outside Gemini 4 by eight minutes, finally ended it with the reluctance of a country boy forsaking a cool creek.

After Colonel Leonov's mission, there was a lapse in Russian manned flights that lasted for more than two years. Gemini, in the meantime, surged through a series of engineering triumphs.

In Gemini 5, Gordon Cooper and Pete Conrad tested a prototype fuel cell that became a vital element in future space flights. Gemini 6 pilots Wally Schirra and Thomas P. Stafford chased Gemini 7's Frank Borman and Jim Lovell through space for 100,000 miles to achieve a historic rendezvous, bringing the two spaceships within a foot of each other.

On the next flight, Neil Armstrong and David Scott took an additional step. After rendezvous with an unmanned Agena target vehicle, Gemini 8 smoothly inserted its nose into Agena's docking cone. Locked together, the two craft flew in tandem for 27 minutes; then a short-circuited Gemini thruster put them into a dangerous spin. Armstrong had to undock, then activate his re-entry system to regain control, thus ending the mission.

Tom Stafford and Eugene A. Cernan refined the techniques in Gemini 9 when they achieved rendezvous three times with their target vehicle. John Young and Michael Collins in Gemini 10 not only latched on to the 26-foot Agena target but also used its 16,000-pound-thrust engine to boost them to a new record apogee of 476 miles. During Gemini 11's first orbit, Pete Conrad and Dick Gordon docked with Agena, employed its engine for a new altitude record of 853 miles, then set a new speed record of 17,897 miles per hour as they looped back to low earth orbit.

Geminis 9, 10, and 11 provided additional space-walk experience through the efforts of Cernan, Collins, and Gordon. The final flight of the series, Gemini 12, became the "catchall." Those experiments most important to the Apollo series went on the schedule, including docking and a space walk by Edwin Aldrin. When Gemini 12 splashed down, Jim Lovell in his two Gemini missions had spent nearly 18 days in orbit.

Side by side with the flight experience of Gemini, the Apollo program began to gain form and substance. One decision alone, on the best way to get to the moon, required more than a million man-hours of technical study. Finally, space officials chose the lunar orbit rendezvous method, rejecting a direct flight as requiring too large a booster for the return trip.

There were unpredictable and tragic delays, however, in the manned missions of both Russia and the U. S.

The United States suffered first, on January 27, 1967: Astronauts Gus Grissom, Ed White, and Roger B. Chaffee were killed when a fire broke out during a routine checkout of their Apollo spacecraft on Cape Kennedy's pad 34.

Less than three months later, Cosmonaut Vladimir M. Komarov was killed. His spacecraft parachute lines fouled during his attempted return to earth after 18 orbits in Soyuz 1.

In the 20 months that followed the 1967 fire, NASA installed flameproof materials inside the

Apollo spacecraft and devised a new hatch that could be opened in seven seconds. The scheduled commander of the first Apollo flight, Wally Schirra, concluded, "Other than putting ejection seats in, I see no faster way of getting out." Scheduled to fly with Schirra were Maj. Donn F. Eisele and civilian R. Walter Cunningham.

On October 11, 1968, America's third-generation, three-passenger spacecraft roared up from the Cape atop a Saturn I-B. It rode a smaller rocket than the huge Saturn V but the space vehicle was still an imposing 20 stories high.

"She's riding like a dream," Schirra reported.

During Apollo 7's 163 earth revolutions, the crew rendezvoused with their own booster and carefully tested all the new systems. On October 22, after a flight of more than four million miles, they splashed down 325 miles south of Bermuda.

So successful had been Apollo's baptism in space that even as my wife Toby and I watched the crew show their films to President and Mrs. Johnson, the rumor at the LBJ Ranch was that Apollo 8, instead of remaining in earth orbit, would be sent into lunar orbit. That decision was announced nine days later in Washington.

Lift-off for Frank Borman, Jim Lovell, and William A. Anders came on the shortest day of the year, December 21, 1968. The great rocket pushed steadily upward on a tail of brilliant flame 700 feet long. An hour and 18 minutes later, in earth orbit, the crew heard the welcome words from communicator Michael Collins, "You are go for TLI [translunar injection]. You are go for the moon"—the brief message that sent man on his longest journey yet.

The magnitude and dramatic symbolism of the Apollo 8 mission became apparent when we were able to see, for the first time, earth as it looks from hundreds of thousands of miles in space. The crew described and brilliantly photographed a cloud-swirled blue, green, and brown sphere serenely sailing through the dark ocean of space.

One of the many gripping moments of the mission was the emergence of the spacecraft in lunar orbit after a 33-minute communications blackout caused by the great shield of the moon—2,160 miles rim to rim.

"O.K., Houston," Lovell called in, immediately getting down to business: "The moon is essentially gray . . . looks like plaster of Paris . . . the walls of the craters are terraced."

The crew then described an aerial journey across the Foaming Sea, past the Sea of Fertility, over the crater Gutenberg, along the Pyrenees Mountains. Men at last were looking at close range upon the moon's snowless mountains, its waterless valleys, its craters and sinuous rilles.

Borman, Lovell, and Anders circled the moon ten times. On the ninth orbit, the bone-tired crew read from the Book of Genesis, and described the beauty of the earth—that lovely home of all mankind. Then, when Christmas Day was only ten minutes old, they fired their main engine behind the moon—again during communications blackout—to start them on their return journey.

The next two Apollo missions were scheduled

NASA

SPACE-WALK MISSION: *Command pilot Maj. James McDivitt (foreground) and Maj. Edward H. White II recline before sealing the hatch of Gemini 4, June 3, 1965. After the launch into earth orbit, McDivitt piloted the two-man craft as White became the first American astronaut to walk in space.*

FLOATING IN SPACE *100 miles up, Major White moves freely during his 20-minute venture from Gemini 4. Tethered by a 30-foot umbilical cord containing oxygen and communication lines, he maneuvered by releasing gaseous nitrogen from a hand-held propulsion gun. Although he traveled some 6,000 miles outside the space capsule at 17,500 miles an hour, White said he "had little sensation of speed and no sensation of falling—only a feeling of accomplishment."*

to check and double-check all the systems, procedures, and maneuvers required for the highly complex lunar landing. Chief among these was the most innovative vehicle ever fashioned on earth: the LM, or lunar module, first passenger vehicle designed to function outside earth's atmosphere and away from earth's gravity.

On earth, the LM looked like a huge awkward insect that had to be tucked and folded inside the rocket. Once in space it would emerge, be turned around and linked in position and finally allowed to unfold its spider-like legs in the sunlight. It had no heat shield and was far too delicate ever to return to earth. But it could take two men down to the moon, and its top portion could then blast upward to bring the crew back for lunar rendezvous and docking. After that—unless in extreme emergency its life-support systems might still be needed—it was useless, and would be jettisoned.

The first LM scheduled to fly was appropriately named *Spider* by the Apollo 9 crew: Jim McDivitt, Dave Scott, and Russell L. Schweickart. Lift-off came on March 3, 1969. In earth orbit the crew fired small thrusters to detach *Gumdrop,* the command and service module, from the rocket's third stage and turn it around; then McDivitt aimed the command ship's ten-inch probe at the cone-shaped receptacle of *Spider,* still snug in its cocoon. After the two craft linked up, McDivitt flipped a switch and springs pushed the combination free of the rocket unit.

Over several days the crew checked the combination's reactions to various rocket burns, tested the lunar suits and their portable life-support systems with a dual space walk, then prepared for the first undocking. With McDivitt and Schweickart in *Spider,* Scott punched the button that separated the two craft. *Spider* maneuvered about a hundred miles away to prepare for the crucial operation of rendezvous and docking. Each of

JAMES A. McDIVITT, NASA

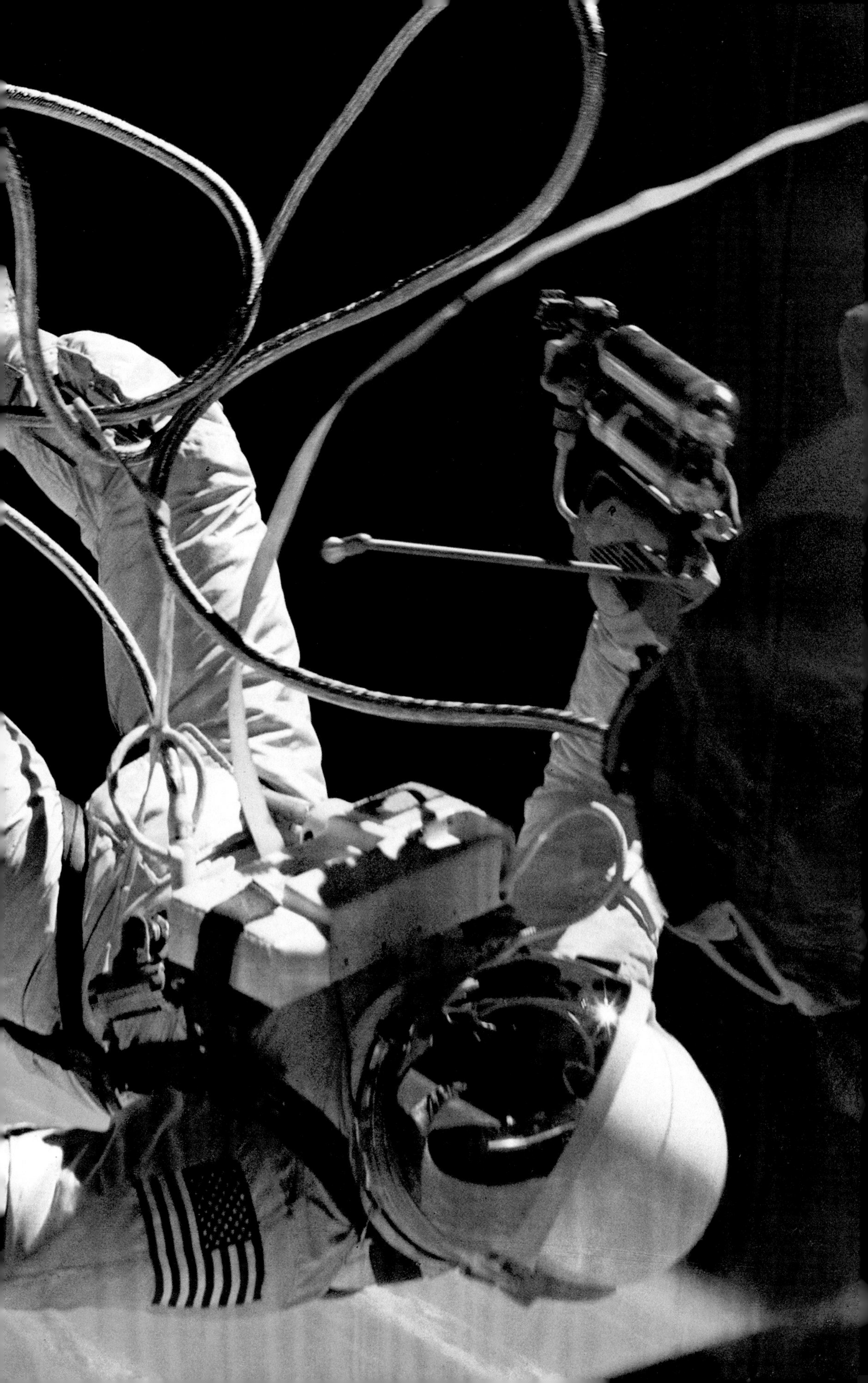

UNITED

NOSE-TO-NOSE *185 miles high, Gemini 7 and Gemini 6 coast 43 feet apart during the world's first space rendezvous, on December 15, 1965. Such experiments helped refine techniques later used on Apollo flights. Scheduled to pilot the first Apollo spacecraft, Astronauts Virgil I. Grissom, Roger B. Chaffee, and Edward White (left to right) died inside their command capsule when fire broke out during a preflight test on January 27, 1967.*

NASA (BELOW); THOMAS P. STAFFORD, GEMINI 7

several engine burns now represented a firing rehearsal for lunar rendezvous and docking, but to McDivitt and Schweickart they were even more important: The burns were their only means of getting back to *Gumdrop* and to earth.

When the two craft were finally in sight of one another, Scott radioed *Spider,* "You are upside down." To McDivitt this was a somewhat unscientific assumption. "One of us isn't right side up," he corrected.

When the docking and crew transfer were completed, *Spider* was jettisoned into a high earth orbit. *Gumdrop* and the crew were plucked out of the Atlantic on March 13.

Spider's performance encouraged arguments by some that Apollo 10 should now attempt the first lunar landing. But more cautious reasoning prevailed. NASA told the veteran Apollo 10 crew, Tom Stafford, John Young, and Eugene Cernan, that they should not attempt to land; but they could go to the moon and take the LM down, down, down to within nine miles of the surface, and scout and photograph the primary Apollo 11 landing zone. That was mission enough.

After launch on May 18, 1969, the command ship *Charlie Brown* with lunar module *Snoopy* in tow coasted toward the moon, relaying vivid television pictures of both planet and satellite. About 60 miles above the forbidding lunarscape, the two craft separated.

Now *Snoopy* was to have a real exploit; just how dangerous it was appears from the fact that only a few extra seconds of thrust during the descent burn could cause *Snoopy* to crash.

Stafford and Cernan knew this, of course, when they fired *Snoopy's* descent rocket, which took them in a long, sweeping arc down toward the surface. Then, as they jettisoned their descent stage, *Snoopy* suddenly pitched, dipped, and bucked—as Cernan said—"all over the skies." Stafford struggled to regain control. Finally the bucking stopped. The difficulty turned out to have been a faulty switch-setting on the instrument panel.

When the reconnaissance was completed and the rendezvous with *Charlie Brown* achieved, *Snoopy's* moment of glory was over, and the crew turned him loose. "He treated us pretty well today," Cernan said as he watched the LM pass out of sight.

Within an hour after the crew of Apollo 10 stepped on the deck of the U.S.S. *Princeton,* NASA Administrator Dr. Thomas Paine announced in Houston, "We will go to the moon. Tom Stafford, John Young, and Gene Cernan have given us the final confidence to make this bold step."

The grave responsibility of the Apollo 11 mission went to civilian Neil Armstrong and two West Pointers, Michael Collins and Edwin Aldrin, both Air Force colonels.

When I reached Cape Kennedy on July 15, 1969, the first arrivals of more than a million visitors that would witness the next morning's launch were already pouring in. Just before dawn, as the press buses inched toward Merritt Island in bumper-to-bumper traffic, people of all ages lined both shoulders of the road.

Not far from pad 39A were assembled more than 4,000 specially invited guests, including former President Johnson and Mrs. Johnson, Vice President and Mrs. Agnew, 19 governors, and 69 representatives of foreign governments.

SOVFOTO (ABOVE AND RIGHT); NOVOSTI

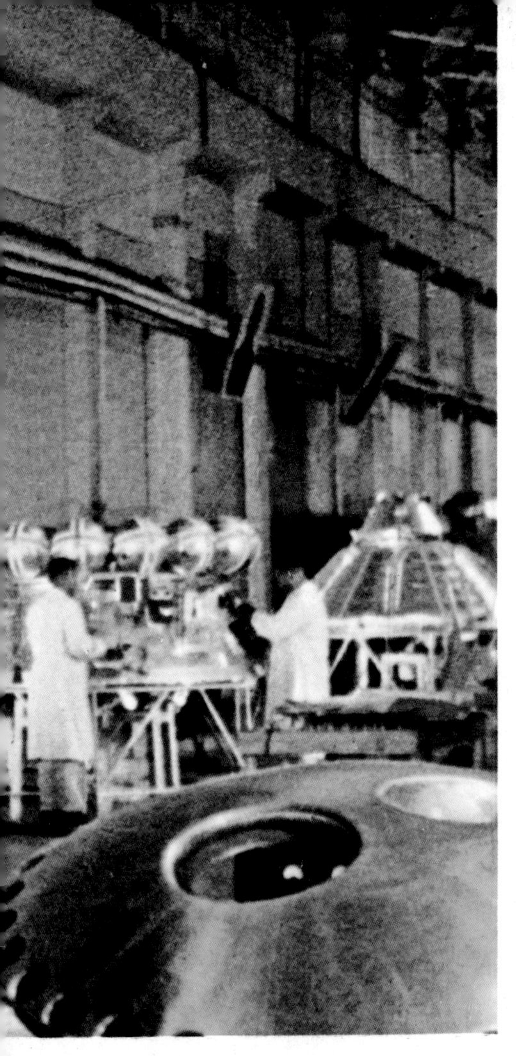

VOSTOK: SPACE VEHICLE FOR FIRST COSMONAUTS

In a Soviet factory, technicians assemble 2½-ton capsules for one-man missions into earth orbit. An escape hatch rests at right. Six cosmonauts, beginning with Yuri Gagarin, rode such spherical spacecraft between 1961 and 1963. In June 1963 auburn-haired Lt. Valentina V. Tereshkova, 26, became the first woman in space when a rocket carried her aloft from the Russian space center at Tyuratam in the Asian Republic of Kazakhstan. At one point of her 49-orbit journey in Vostok 6, she passed within three miles of Vostok 5, piloted by fellow cosmonaut Lt. Col. Valeri F. Bykovsky. At left, Lt. Tereshkova sorts her gear after landing on a plain in Kazakhstan. At right, braces fall away from a Vostok rocket lifting off at Tyuratam.

The great moon rocket, as Dr. Robert Goddard and Konstantin Tsiolkovsky had long ago conceived, was in fact a "rocket train." And its astro-conductors were aboard and ready.

The target was an oval-shaped sector of the Sea of Tranquillity 218,096 miles away, a dry, stark, airless, lifeless desert.

In the final few seconds before launch, two white pelicans soared across my view, reminding me that I had fished here before the wild shoreline became the free world's first spaceport. Now, for the entire planet, this spot was land's end.

The great two-pronged flame blossomed. In awesome silence the journey began. I waited for the shock wave, and glanced again at the two pelicans to see their wings suddenly raked and shaken by the invisible force. Startled, they dived off in a long glide toward Pelican Island.

Land's end erupted. The deafening staccato of what Norman Mailer called "a ship of flames" smote lung cavities, sinuses, bone marrow—"and the earth began to shake and would not stop."

Finally we could hear Armstrong's amplified voice from the speakers: "In-board cut-off." In six minutes, all that was left of the volcanic violence of the launch was a giant condensation cloud drifting above the singed and smoking mobile launcher.

During the drama of such a magnificent event it was inevitable that—along with the prayers,

NASA (BELOW); ROCKETDYNE DIVISION, NORTH AMERICAN ROCKWELL

POWER FOR APOLLO, *five J-2 engines (right) cluster at the base of the second stage of a Saturn V rocket. Igniting some 40 miles up after first-stage burnout, the liquid-propellant J-2's push the moon rocket toward earth orbit. The third-stage engine provides the thrust needed to escape earth's gravity. Above, lunar module* Eagle *gets a final systems check at Cape Kennedy before beginning its historic journey to the moon.*

cheers, and shouts—there would be tears of joy or relief or pride. A few of these, I noticed, coursed down the cheeks of the dedicated chief researcher of this book, Johanna Farren.

Another witness was R. Sargent Shriver, Jr., then U. S. Ambassador to France, who afterward recalled something said to him by the man for whom the launch site was named. Concerning the decision to send men to the moon, President Kennedy remarked:

"I firmly expect this commitment to be kept. And if I die before it is, all of you here now just remember when it happens I will be sitting up there in Heaven in a rocking chair just like this one, and I'll have a better view of it than anybody."

The three-man expedition entered lunar orbit on the afternoon of July 19. The next day, after Armstrong and Aldrin separated the lunar module *Eagle* from the command ship *Columbia,* Collins radioed, "The *Eagle* has wings." At 3:10 p.m. the *Eagle* crew heard the command they were waiting for: "You are go for DOI [descent orbit insertion]."

The crew was aware by this time that a mysterious Soviet craft, Luna 15, was also in lunar orbit. Was it possible that the Soviets planned to precede Apollo 11 to the moon's surface?

If the men felt any apprehension over the presence of the Soviet vehicle, it was not apparent in the suspenseful final minutes of *Eagle's* descent. Armstrong suddenly realized they were coming

NASA

NEXT STEP: THE MOON. *Command pilot Neil Armstrong, with an air conditioner to cool his pressure suit, crosses a 320-foot-high swinging bridge from the launch tower to the Apollo 11 spacecraft. Michael Collins and a technician follow.*

down toward a depression littered with large boulders. He took over from the automatic control system and gingerly guided *Eagle* across a crater the size of a football field.

Aldrin's voice came clearly out of the ceiling in the Houston press center: "... got the shadow out there ... 75 feet, things looking good ... picking up some dust ... 30 feet ... drifting to the right a little. ..." They were nearly out of fuel.

A few seconds later, a 68-inch probe extending down from the LM touched the moon, turning on a contact light on the instrument panel. Instantly, Armstrong shut off the engine. They felt a jolt like that of a jet touching down on a runway.

"Houston, Tranquillity Base here," Armstrong finally reported. "The *Eagle* has landed."

The next step was to walk on the lunar plain and pick up priceless pieces of pay dirt, but it took a while to get ready—about 6½ hours of scheduled rest and preparation.

As Armstrong, the first lunar prospector, descended the ladder, he pulled a lanyard that unfolded and positioned a television camera. The world thus saw him slowly make his way down to the surface and heard him describe his progress.

When Buzz Aldrin joined him on the powdery, somewhat slippery surface, Armstrong read from a 9-by-7⅝-inch plaque attached to *Eagle:*

"Here men from the Planet Earth first set foot upon the Moon July 1969, A.D. We came in peace for all mankind."

Aldrin next danced a jig—in the interest of science. Walking, running, hopping, he cavorted in front of the television camera, half-floating like a marionette. During his panting commentary he explained, "You do have to be rather careful to keep track of where your center of mass is. ... like a football player, you just have to cut out to the side. ..."

After collecting preliminary lunar samples, the astronauts were planting the U. S. flag when they were interrupted by the first earth-moon telephone call. It was from the White House. President Nixon praised the men for their achievement and added, "For one priceless moment in the whole history of man, all the people on this earth are truly one. ..."

Ceremonies over, the prospectors quickly began work. They deployed the solar-wind experiment and a seismometer to register quakes and meteorite impacts, set up a laser-beam reflector for precise measurement of the moon's distance from earth, and collected 47.8 pounds of samples.

At the end of 2 hours, 31 minutes, and 40 seconds of carefully planned exploration, the rock boxes, solar-wind package, and film had been stowed and the astronauts had reboarded. Exhausted, they rested on *Eagle's* metal floor. The last of their well-earned nap came on Monday, July 21, a national holiday proclaimed in their honor by President Nixon.

Seven hours later, they heard Mission Control advise, "... you're cleared for take-off."

For the second time there was fire on the moon, as the upper half of *Eagle* lifted off. When the LM redocked 3 hours and 39 minutes later with what Mike Collins called his "mini-cathedral," they learned that the unmanned Soviet Luna 15 had crashed in the Sea of Crises, some 500 miles east of Tranquillity Base.

During the 60-hour return to earth, as the three voyagers restored their energies, writers, philosophers, politicians, and scientists struggled to interpret and signify the still incomprehensible achievement. Wrote poet James Dickey:

We will take back the very stones
Of Time, and build it where we live. . . .

Reconstructing time from the calendar of the moon was uppermost in Neil Armstrong's mind when he radioed capsule communicator Charles M. Duke, Jr., "Open up the LRL doors, Charlie."

LRL was the $8,500,000 Lunar Receiving Laboratory at Houston where the crew was to be quarantined and the priceless cargo analyzed. The purpose was to ensure that nothing alien could contaminate earth. Before the flight, Dr. Persa Raymond Bell, then LRL manager, had told me, "We don't really expect to find dangerous organisms in the lunar samples, but you must

APOLLO 11 *roars toward the moon on July 16, 1969. As the mighty Saturn V rocket lifted skyward, thousands of spectators cheered, cried, laughed, or stood in silent awe. Many, like the woman below, kept their fingers crossed. Neil Armstrong summed up the lift-off in an early report to Mission Control: "Saturn gave us a magnificent ride."*

THOMAS R. SMITH, N.G.S. STAFF

PORTRAIT OF A LUNAR VOYAGE, *painted before the first Apollo mission, accurately projected the flight of Apollo 11. The Saturn V rocket's second-stage engines ignite (1) following first-stage burnout and separation 162 seconds after lift-off from Cape Kennedy on July 16, 1969. The second stage accelerates Apollo to a speed of 15,500 miles an hour, then falls away as the third stage (2) pushes the spacecraft into parking orbit around the earth. After circling the globe one and one-half times the third-stage engine fires once more (3), hurling the payload on a course for the moon. Half an hour later, four adapter panels open, and the command and service module* Columbia *separates and begins pulling away (4) for an on-course docking operation with the lunar module* Eagle. *After moving about 100 feet out,* Columbia *turns around (5) and docks gently with the buglike craft.*

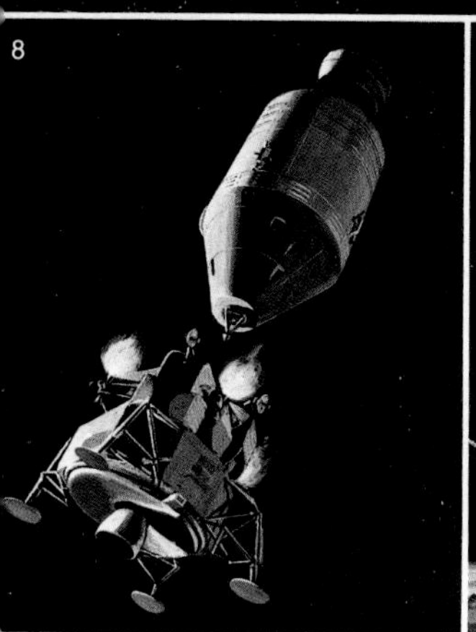

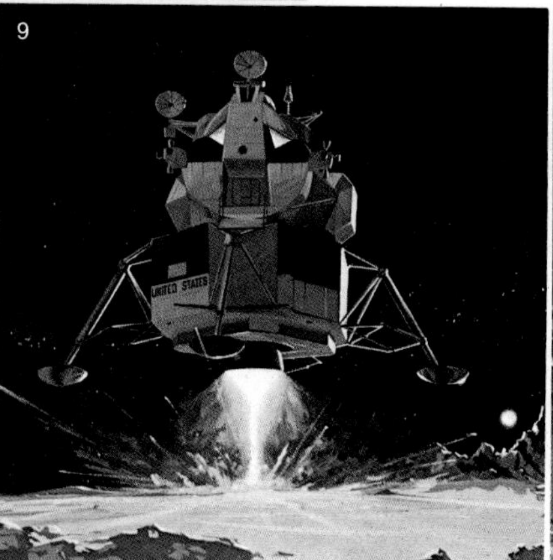

remember that this port of entry has responsibility for our entire planet. We can't be too careful."

After splashdown on July 24 and decontamination of *Columbia,* the three astronauts went immediately into a small stainless-steel mobile home aboard the U.S.S. *Hornet.* President Nixon was present to welcome them; then medical personnel inside their Mobile Quarantine Facility began examinations even as the *Hornet* headed for Hawaii. From there a C-141 cargo plane flew the unit and its occupants directly to Houston.

The precious lunar samples had already arrived in two vacuum-sealed boxes transported in separate planes. Inside the receiving lab, the boxes were bathed in ultraviolet light and cleansed with acid. The actual opening of the treasure chests was begun in a vacuum chamber by technicians reaching in with special gloves.

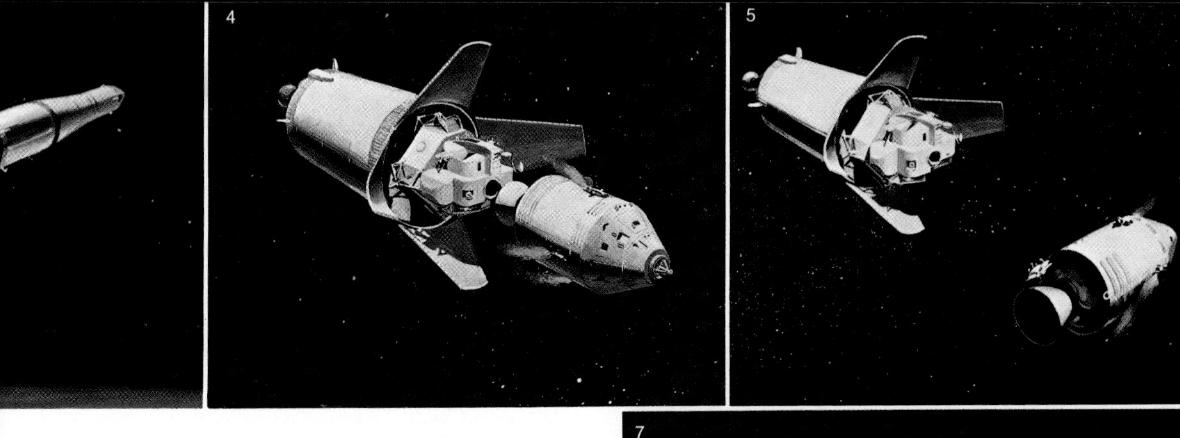

Springs eject the linked vehicles from the spent third stage (6). Columbia's *rockets then fire again, and the craft speeds away (7). Four days later, on July 20, the moon looms large in* Columbia's *hatch windows. Neil Armstrong and Edwin Aldrin crawl into* Eagle *and power it up. The two craft undock and draw apart (8); while Michael Collins holds* Columbia *in lunar orbit,* Eagle's *descent engine eases it to touchdown (9). Next day the first two men to stand on the lunar surface fire* Eagle's *ascent engine (10), using the descent stage as a launch platform. The two craft redock in lunar orbit (11), and Armstrong and Aldrin rejoin Collins aboard* Columbia. *Several revolutions later the craft jettisons* Eagle *(12) and remains in lunar orbit. Ready for the return to earth, the astronauts fire* Columbia's *engine (13), sending the craft toward its splashdown in the Pacific Ocean on July 24.*

NORTH AMERICAN ROCKWELL, EXCEPT (9) TELEDYNE RYAN AERONAUTICAL

At first glance scientists were disappointed; gray dust entirely concealed the rock surfaces. After cleaning, however, the rocks initially offered both surprises and mysteries. For one thing, there was far more glass than anticipated. Minerals identified early included remarkably high percentages of titanium (relatively rare on earth), along with feldspar and olivine. Later, traces of carbon were found, but a 200,000-power electron microscope revealed no hint of any life form or fossil.

Dust samples inserted under the skin of mice, quail, and insects had no apparent effect. When a pinch of moon dust was heated to 3,000° F., an analysis of the escaping vapors revealed such rare gases as argon, neon, krypton, and xenon, all constituents of the solar wind.

What did it all mean in terms of the age and origin of the moon and solar system? The first

theories were as tentative as those of a visitor to earth doubtless would be had he only 47 pounds of South Dakota to analyze.

One of the most significant early discoveries resulting from Apollo 11 came in late August when dating processes in the LRL suggested the age of most of the moon rocks to be 3.3 billion years or more, according to Dr. Wilmot Hess.

In September portions of the lunar pay dirt were released to 142 experimenters in many parts of the world, and they met in January 1970 to discuss their tentative conclusions: Although there was no evidence of life—or, at that point, water—on the moon, the basic chemistry of the satellite is similar to that of earth. Many of the Apollo 11 rocks are volcanic, but clearly distinctive from earth rocks. All of earth's elements apparently exist on the moon, but some have combined into minerals unknown on earth.

Said Pete Conrad before his moon flight with Dick Gordon and Alan Bean, "the name of the game of Apollo 12 is lunar surface exploration."

He was referring to the fact that on Apollo 11 other duties had restricted the actual sample-gathering time to slightly more than 60 mintues. Detailed documentation, so important to astrogeologists, had been at a minimum. Apollo 12 was scheduled for two periods of extravehicular activity totaling almost eight hours.

On November 14, 1969, the all-Navy crew

NASA (BELOW); TOR EIGELAND, BLACK STAR

ARABS IN KUWAIT *gather before a television set during the Apollo 11 mission to witness man's farthest journey. An estimated five hundred million people—a seventh of the population of earth—watched Astronaut Neil Armstrong as he took his "small step" (above), televised to earth from a camera mounted on the* Eagle's *descent stage.*

blasted upward in the command module *Yankee Clipper* with LM *Intrepid* in tow. About 36 seconds after launch into a dark cloud, the rocket became the first manned space vehicle known to be struck by lightning. So many red warning signals came on that the crew couldn't read them all.

When no permanent damage could be detected, they were given a "go" signal for the moon. Their prime landing zone was beside the unnamed crater where the 625-pound Surveyor 3 had landed in April 1967. Before the flight, the crew estimated they had only a 50-50 chance of landing within walking distance. Yet on November 19, as Conrad stepped out on the Ocean of Storms, he exclaimed to Bean, "Look what I see sitting on the side of the crater. Surveyor! . . . It can't be any farther than 600 feet from here. How about that?"

Unfortunately the television camera went blank, but the two men's running commentary kept their earth audience well posted as they set up the equipment of the $25,000,000 ALSEP (Apollo lunar surface experiments package). After collecting and documenting 13.7 pounds of samples, they rested inside *Intrepid* for eight hours, then emerged again to visit Surveyor 3. They found the three-legged robot covered with dust, apparently kicked up by their landing. They sawed off portions, including the 17-pound television camera, for analysis back on earth and then reboarded *Intrepid* for the moon's second spacecraft launch.

SOYUZ 10 *rests on its side before launch at the Tyuratam space center. Built and tested in a horizontal position, the spacecraft rides standard rail equipment to the launch pad. There a hydraulic lifter raises it upright, and arms of the launch tower close around it like the petals of a flower. The covered stairways (right foreground) lead to banks of instruments that monitored the lift-off. On April 24, 1971, the craft docked with Salyut, a research station already in orbit. After flying together for 5½ hours, the craft separated. Then Soyuz 10 returned Cosmonauts Vladimir A. Shatalov, Alexei S. Yeliseyev, and Nikolai N. Rukavishnikov safely to earth in the Soviet Union's first night landing.*

After link-up with Dick Gordon in *Yankee Clipper,* they sent the now expendable LM on a crash course toward the moon—and another mystery developed. Houston flight controllers counted down for the impact, then were startled to discover that the seismometer left on the surface continued to register vibrations far longer than could have been duplicated anywhere on earth.

Said Dr. Maurice W. Ewing of the Lamont-Doherty Observatory, "It was as if one had struck a bell a single blow and found that the reverberations continued for 55 minutes."

NOVOSTI

Since Apollo 12 and Apollo 11 had landed on different parts of the moon, scientists were unwilling to risk earth contamination and again insisted on a quarantine. The 75.7 pounds of Apollo 12 samples showed numerous differences from the earlier rocks. These were slightly lighter in color and generally larger (the one Bean called a "grapefruit" weighed four pounds). As in the Tranquillity samples there was metallic iron, which is very rare in earth rocks; on the other hand they contained much less glass than the Apollo 11 samples. Among the most intriguing was the now famous granddaddy rock, believed to be more than four billion years old, older than anything known on earth.

"This means," said Dr. Robin Brett, chief of the Geochemistry Branch at the Manned Spacecraft Center, "that there may be rocks on the moon that have been unchanged since they first crystallized soon after the solar system was formed. Since all such rocks on earth have long since been erased, our ability to look back in time will now increase and help us better understand the early history of the earth."

TRIUMPH AND TRAGEDY: *Russia's Soyuz 11 approaches the Salyut space station (foreground) on June 7, 1971, in this artist's conception. After docking, the three cosmonauts below entered the boxcar-length laboratory, sent up 49 days before. From left, during training: Lt. Col. Georgi T. Dobrovolsky, commander; Victor I. Patsayev, test engineer; Vladislav N. Volkov, flight engineer. After their return from a record, highly successful 24-day mission, ground crewmen opened the hatch—and found them dead. A leak in the hatch's seal had caused a rapid, fatal decompression.*

Additional data came from the magnetometer left on the surface, which showed the moon has a magnetic field far stronger than expected. The portions of Surveyor 3 showed negligible meteoroid damage but a sandblasted effect, presumably from the dust shower of the LM landing.

With two visits having been made to the lunar flatlands, or maria, scientists now wanted samples and data from one of the lighter-colored uplands of the moon. The site selected for the Apollo 13 mission was below the lunar equator in the hilly area near the crater Fra Mauro.

Scheduled to undertake the difficult landing in *Aquarius* were Jim Lovell, the dean of space pilots, who on three missions had logged 572 hours in space; Fred W. Haise, Jr.; and Thomas K. Mattingly II. The weekend before the flight Mattingly was exposed to German measles and was replaced by backup crewman John L. Swigert, Jr.

For two days after lift-off on April 11, 1970, everything went well. Then at 10:08 p.m. EST, when *Odyssey* was 205,000 miles from earth and Lovell had just wished earthmen "good night," there was a dull explosion.

It quickly became apparent that *Odyssey* had suffered a serious loss of power. Soon the crew reported a vapor leak. Lovell suspected a loss of two of the power-producing fuel cells in the service module. When Mission Control ordered him to power down, the cabin grew so dark the men got out flashlights. Within 90 minutes, one oxygen pressure gauge registered zero and the other was dropping.

Then the ground advised ominously, "We

NOVOSTI (ABOVE); SOVFOTO

figure we've got about 15 minutes' worth of power left in the Command Module. So we want you to start getting over in the LM and getting some power on that."

The only possible lifeboat was the "aluminum balloon" of the LM. In it the men would loop all the way around the moon before starting home.

Lovell and Haise crawled through the tunnel and powered up the LM's relatively weak batteries and limited life-support systems, rigging a hose from a spare space suit to relay oxygen to Swigert, who wanted to keep an eye on *Odyssey*.

The prime objective of the tense experts at Mission Control was to get the spacecraft back on a trajectory that would bring it in over any of the earth's oceans. This involved, first of all, an attempt to fire LM's descent engine to put the craft on a course for splashdown in the Indian Ocean. A later LM burn attempted to accomplish two purposes: to speed up the return and to shift the landing zone to the Pacific.

With the moon landing canceled and the goal now simply to get home, the mission still made a significant contribution to lunar science. Just before the spacemen passed behind the moon, their third-stage rocket crashed on the surface. When Mission Control reported to *Odyssey* that "your booster just hit the moon, and it's rocking a little bit," Lovell replied, "Well, at least something worked on this flight." The 15-ton rocket hit just 85 miles from the Apollo 12 seismometer and caused the moon to ring for about four hours.

During the precarious journey back to earth, ingenious improvisations kept life-support units

RICHARD F. GORDON, JR., NASA

HEADING FOR THE OCEAN OF STORMS TO WIDEN MAN'S HORIZONS

Second moon landing begins as the Apollo 12 lunar module Intrepid *descends toward pitted lowlands on November 19, 1969. Capt. Charles Conrad, spacecraft commander, and Capt. Alan Bean, lunar module pilot, made two excursions totaling 7 hours and 29 minutes, to gather rock samples and deploy a nuclear-powered experiments package. Capt. Richard F. Gordon, Jr., pilot of the command module* Yankee Clipper, *took photographs of the lunar surface from orbit, concentrating on features selected for study by earthbound scientists. At right, sunrise sweeping across the moon's far side highlights the rim of the crater Gambart (top).*

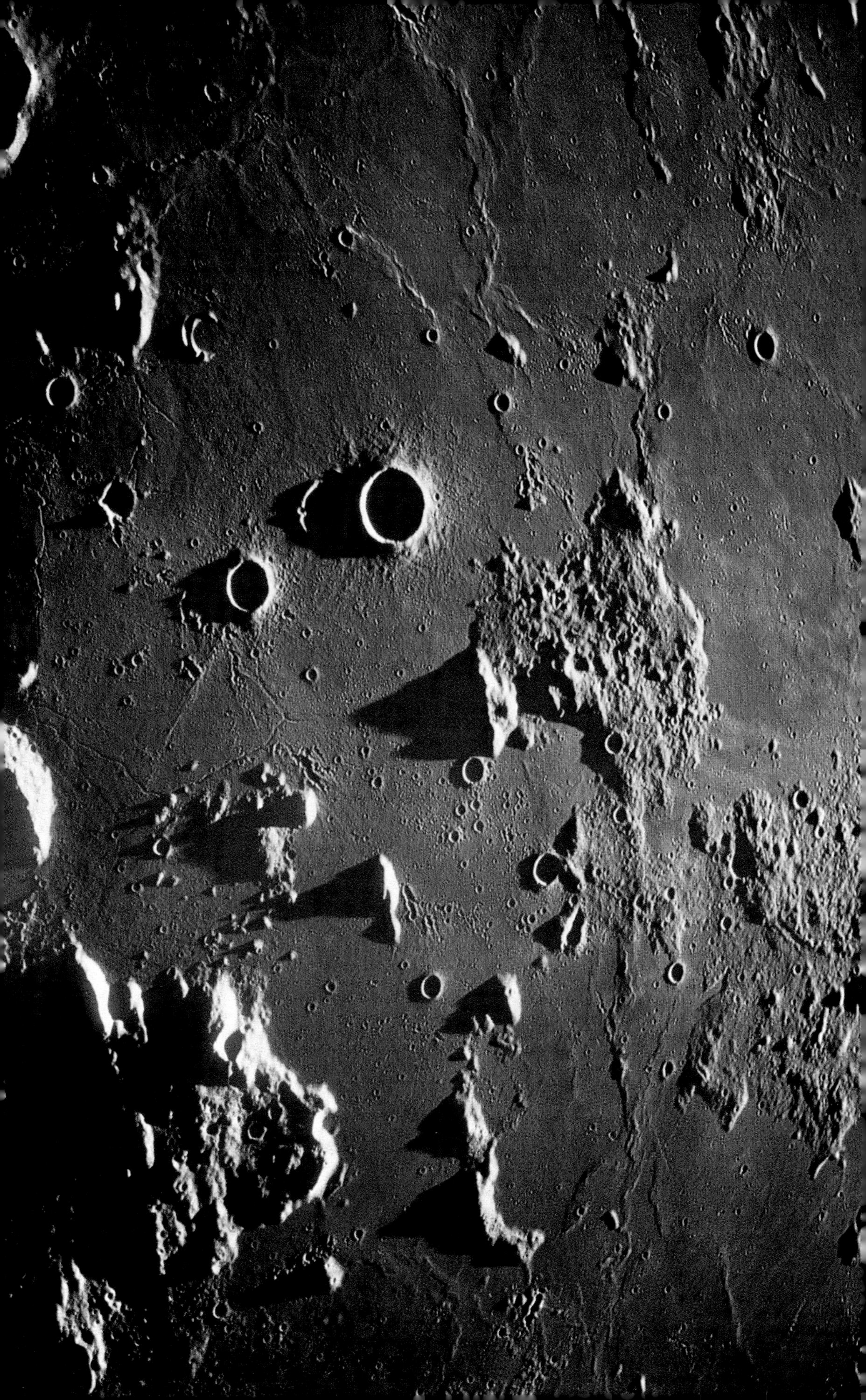

NEIL A. ARMSTRONG, NASA

going, although the temperature dropped to 38° F. and the windows were partly frosted over. The return leg was just under way when refined data showed the present course would miss the earth by 99 miles. Still another burn, the largest in-course correction yet attempted, was required. It worked. The spacecraft *Odyssey* and the carrier U.S.S. *Iwo Jima* were finally heading for the same spot in the Pacific.

Before reaching earth's atmosphere, the shivering crew of Apollo 13 released and photographed the badly damaged service module, and, finally, the LM lifeboat that had saved their lives.

"Farewell, *Aquarius*," said Mission Control, "and we thank you."

"She sure was a great ship," was Lovell's salute.

After the 3½-day ordeal, millions around the world reacted with relief and gratitude as the three great orange-and-white parachutes appeared on television screens. Lovell, Haise, and Swigert splashed down just four miles from the *Iwo Jima*.

The explosion was eventually traced to the failure of a thermostat, which caused overheating of an oxygen tank.

The Fra Mauro uplands remained the target for the revamped Apollo 14. Chosen as commander

MOONQUAKE DETECTOR—*a solar-powered seismometer—stands ready to record tremors. After setting it up, Apollo 11 Astronaut Aldrin returns to* Eagle. *Sensitive enough to register a pea-size meteorite hitting the moon half a mile away, the device transmitted data for five weeks. Beyond Aldrin rests the laser reflector used to measure earth-moon distances precisely; a television camera juts above the horizon to the left of the flag.*

was the first American in space, Alan Shepard. Pilot of the command module *Kitty Hawk* was Stuart A. Roosa; scheduled to walk on the moon with Shepard was Edgar Mitchell.

After launch on January 31, 1971, the crew reached the moon and received a "go" to descend to Fra Mauro. Shepard put the lunar module *Antares* down on target. In their first EVA, Shepard and Mitchell set up the most sophisticated lunar scientific station yet deployed, and collected 41.6 pounds of samples.

At the end of a 6½-hour rest period the astronauts, with their two-wheeled equipment-carrying "ricksha," set out for the rim of Cone crater, nearly a mile away and 330 feet higher than the landing site. The color TV camera clearly transmitted the wheel tracks—the moon's first road.

The slope of Cone crater turned out to be steeper than expected and studded with weirdly shaped boulders. During the strenuous climb, the two men's heart rates increased by some 60 beats per minute. Finally, with EVA time running out, Mission Control ordered them back short of the rim.

Before entering *Antares* for the last time, with 53 more pounds of samples, Shepard took a brief unscheduled recess. Removing the head of a golf club from his suit, he attached it to the handle of one of the tools. He then dropped three golf balls to the surface. After missing with his first swing, he connected, and gloated, "There it goes, miles and miles and miles."

Results of the flight were gratifying. The LM's impact on the moon, this time recorded by two seismometers, triggered signals that continued for about three hours. Most of the other experiments continued to yield data, including a stronger measurement of magnetism at Fra Mauro than at the Apollo 12 site. From *Kitty Hawk* Roosa had photographed possible future landing sites.

The Apollo 14 samples brought into the final quarantine at the Lunar Receiving Laboratory included several football-size rocks, and other rocks and dust that were both lighter in color and distinctly different chemically from previous samples. Most were higher in aluminum and sodium, lower in iron, higher in radioactivity.

Six months after the flight, data released by Dr. G. J. Wasserburg at California Institute of Technology produced another surprise. The rocks gathered during the Apollo 14 mission, generally predicted to be much older than previous samples, turned out to be only about 3.9 billion years old. If the planets and solar system were formed 4.6 billion years ago, as many scientists believe, then

HADLEY BASE, *August 1, 1971: Apollo 15 lunar module pilot James B. Irwin salutes the United States flag. Moon dust smudges his heat-reflecting space suit. Behind him stands the lunar module*

DAVID R. SCOTT, NASA

Falcon; a television camera and an umbrella-shaped antenna aboard the battery-powered moon car Rover (right) beamed the explorations of Irwin and mission commander David Scott to earth.

THE LUNAR VOYAGES: MAN SEARCHES FOR CLUES TO COSMIC MYSTERIES

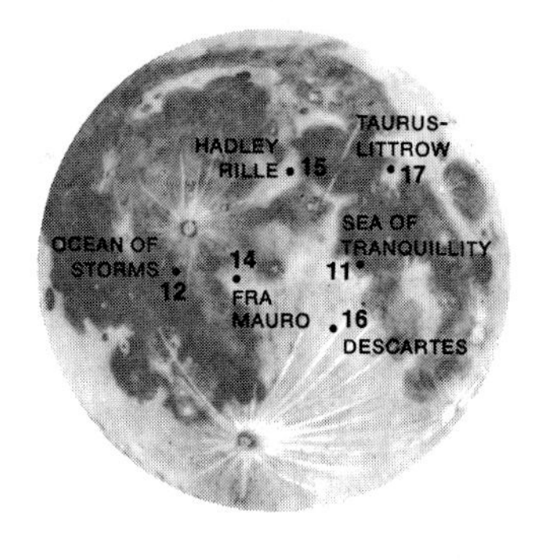

Since man first lifted a rock from the moon's surface, he has measured, probed, and analyzed a growing supply of lunar material—along with a wealth of other data—seeking answers to the ancient questions of how earth's neighbor got there, and when. With such answers come clues to the origin of the solar system.

JULY 1969: On the first lunar mission, Edwin Aldrin set up a solar wind collector (left), a deceptively simple device of thin aluminum that he could roll up and down like a window shade. Exposed for 77 minutes, the foil trapped atomic particles from the solar wind that continually bombards the moon. Brought back to earth, then melted and vaporized, the foil yielded the first reliable measure of the composition of the solar wind, which never reaches earth-based instruments because of the planet's magnetic field.

GEOLOGISTS, geophysicists, geochemists, biologists, astronomers—all helped in choosing areas (above) for the Apollo missions. For some experiments, they needed a triangular network of geophysical instruments. Apollo seismometers, deployed at landing sites a known distance apart, transmit to earth information that shows the intensity and location of seismic activity, however slight. A seismometer at Tranquillity Base recorded even the astronauts' footfalls.

Signals from lunar seismic occurrences differ from any observed on earth. Varying in intensity, they begin weakly, then grow stronger, indicating a scattering of energy by the fractured, rubble-strewn crust. Vibrations detected by the Apollo 12 seismometer average one every day and a half, whereas on earth several hundred occur daily. Scientists attribute the scarcity and weakness of the signals to the internal structure of the moon. With the moon's closest approach to earth each month, ho ever, lunar disturbances increase the surface actually bulges sligh toward earth. This phenomenon lea scientists to ponder whether t moon at the same time can influen perhaps even cause, quakes on ear

THREE APOLLO MISSIONS, 14, and 15, carried laser reflecto among the lunar module's equi ment (below). These devices—re tangular frames holding 100 or mo glass prisms—enable scientists "shoot" at the moon, firing puls of ruby laser light. By clocking t

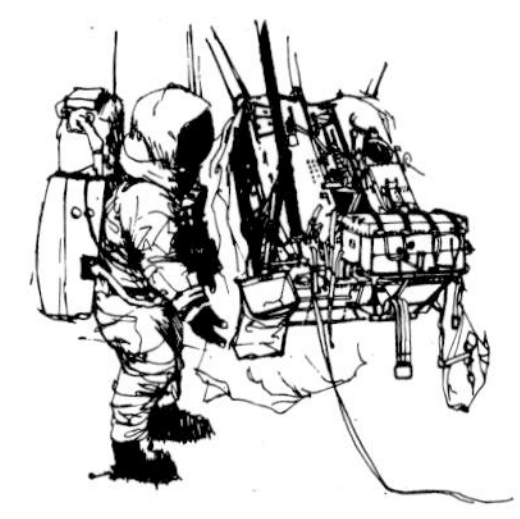

time required for the light to rea the prisms and reflect back, they d termine the distance between ear and moon to within inches.

HELPED by the weak lunar gravi an astronaut (above right) lifts bu equipment, using the antenna ma from the central power station (rig center) as a carry bar. On the moor 180-pound instrument weighs or 30 pounds. To safeguard the expe ments from any contaminating du kicked up by the departing rock

the Sea of Rains must have been formed more than 600 million years later. This meant that large chunks of debris were still floating around the solar system in unstable orbits much later than anyone had thought.

Hadley-Apennine! When the Apollo 15 landing area was announced, a tingle of anticipation ran through every geologist who knew of this dramatic region north of the moon's equator. Hadley Rille is a sinuous chasm 1,200 feet deep and a mile wide; the solid escarpment of the Apennines towers 15,000 feet. What better place for a study of lunar strata? But this juxtaposition of harsh canyon and steep mountain called for a ridge-hopping descent that on earth would have challenged a helicopter pilot.

When my wife and I met with Apollo 15 crew members before the flight, there was something special in the morale and anticipation of mission commander Dave Scott, command module pilot Alfred Worden, and lunar module pilot James Irwin. It was obvious that their primary interest was not the flight itself. They were more excited about the potential yield from the combination of a highly promising site and the most intensive training in geology any Apollo crew had yet received.

At a press conference Scott explained that since Capt. James Cook in 1768 commanded the "first

he astronauts placed them 300 to 00 feet from the lunar module.

POLLO 15 included a special task or David Scott, who drilled out an ight-foot core sample showing 58 istinct layers (far right). On April 0, 1972, John W. Young and Charles . Duke, Jr., landed their Apollo 16 unar module *Orion* in the Descartes egion. In highlands near the equator hey gathered 213 pounds of lunar naterial. Apollo 17 Astronauts Eu- jene A. Cernan and Harrison H. chmitt collected 243 pounds of amples from the Taurus-Littrow alley, which they explored on De- ember 11-13, 1972. The six lunar rews brought back more than 840 ounds of rocks and soil. With such naterial, with seismic and geochem- cal analyses, and with the torrents of data still flowing to earth from ensors deployed by Apollo crew- nen, scientists increase their under- standing of the moon's structure and omposition.

ESPITE SIMILARITIES, the moon iffers greatly from earth. Both are hot internally and divided into layers. Yet measurements place the density of the moon at only three-fifths that of earth. Its radius spans only 1,080 miles—a fourth of earth's. A tenuous lunar atmosphere and the absence of water indicate that the moon never had the kinds of gases that formed the earth's atmosphere and oceans, or that the gases escaped because of weak lunar gravity. New data show that the moon's crust extends almost four times deeper than earth's and presents a lifeless panorama of craters, fractures, and ancient lava beds. Although the same chemical elements make up both bodies, they do so in different proportions. Seismic soundings suggest a partially molten lunar center, but

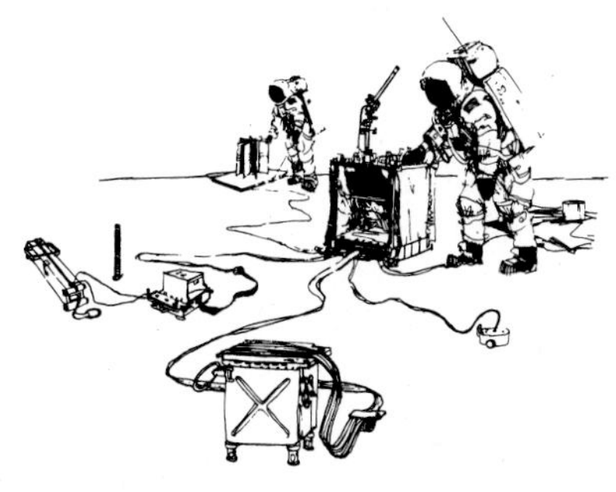

the material does not resemble earth's nickel-iron core.

What of the moon's origin? Did it come from a different part of the solar system? Was it born from the earth, or at the same time from the same cloud of gas and dust? No conclusive evidence to prove any of these theories exists.

Geologists now believe that the moon formed when the earth did—about four-and-a-half billion years ago. But the oldest rocks ever found on earth date back only 3.75 billion years, leaving a three-quarter-billion-year gap in man's knowledge of terrestrial history. Oceans, winds, rains, mountain building, and continental drift have destroyed the early clues. The moon, little disturbed by erosion, has yielded material 4.6 billion years old—a four-gram fragment brought back by the astronauts of Apollo 17, the final manned lunar mission scheduled by the United States.

SCIENTISTS will need years to assimilate and distill the abundance of information gained from the lunar voyages. One geochemist said that data from Apollo 15 alone would keep him and his colleagues busy for five years. Although much about the moon remains an enigma, continuing research will add still further dimensions to the new image of the moon and to man's understanding of the solar system.

purely scientific expedition" in the *Endeavour,* the Apollo 15 crew had decided on that name for their command and service module. The LM's name was *Falcon,* for the mascot of U. S. airmen —unanimous choice of the all-Air Force crew.

Enthusiasm for the Apollo 15 mission spread to a public that appeared to have lost interest in space exploration. At Cape Kennedy the greatest crowd since Apollo 11 was staking out motel and camping space for the lift-off on July 26, 1971.

At 9:34 a.m. the rocket bored aloft through nearly cloudless skies, carrying the new and heavier Apollo "J" spacecraft. A major addition to the cargo was the 460-pound LRV—lunar roving vehicle—folded into the LM like a Murphy bed. When *Falcon* skimmed over a 10,000-foot-high section of the Apennine Mountains and landed on July 30, it was within 1,800 feet of its target.

Early the next earth-morning, Scott and Irwin became the seventh and eighth astronauts to step into the dazzling light of the lunar day. They drove the Rover like a dune buggy on undulating desert. It bounced hard enough to unseat its passengers had they not been belted in place.

". . . boy, look at that!" Scott exclaimed.

". . . you better watch the road," responded the moon's first backseat driver.

At the edge of Hadley Rille the vehicle-mounted

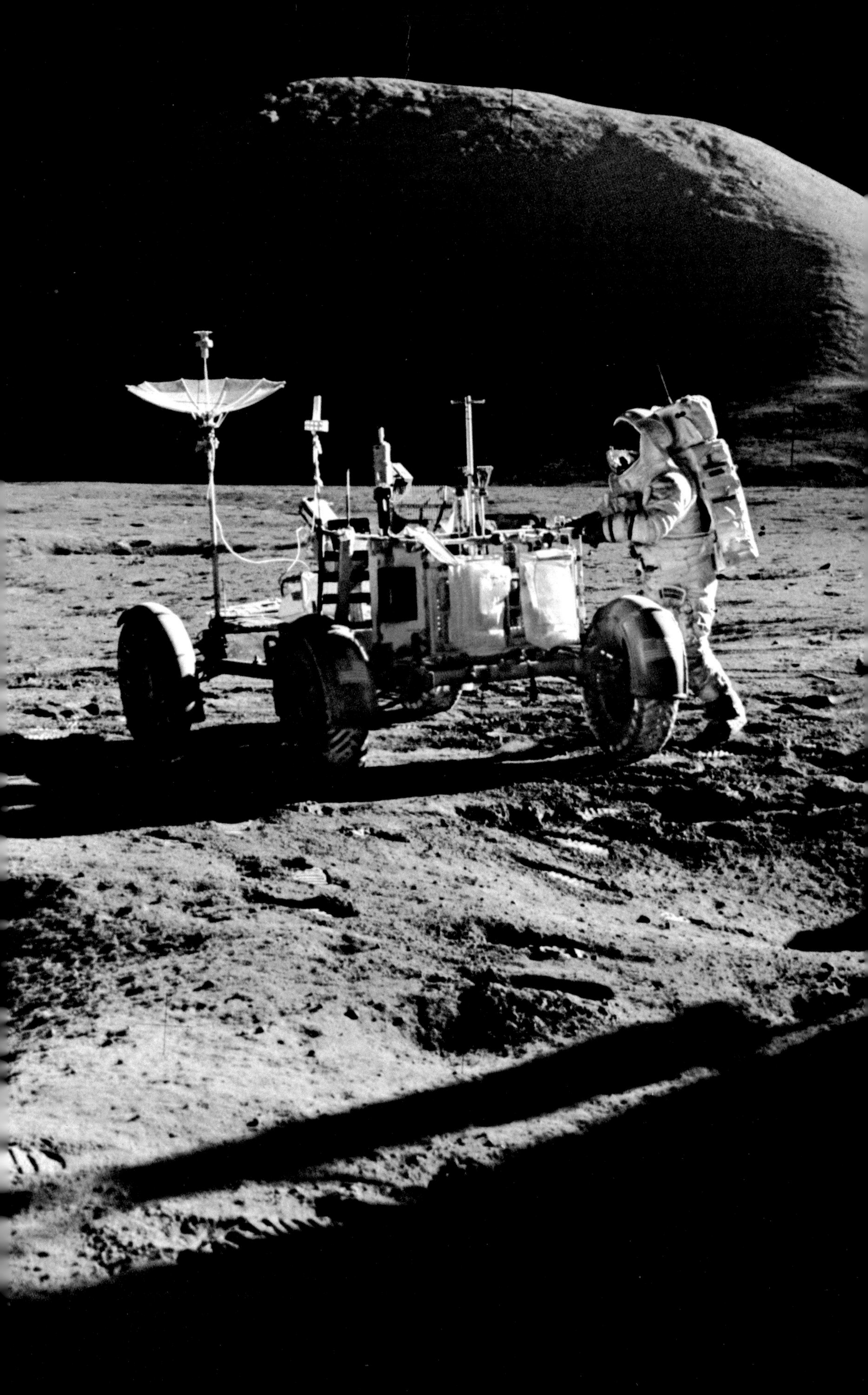

GEOGRAPHIC ART

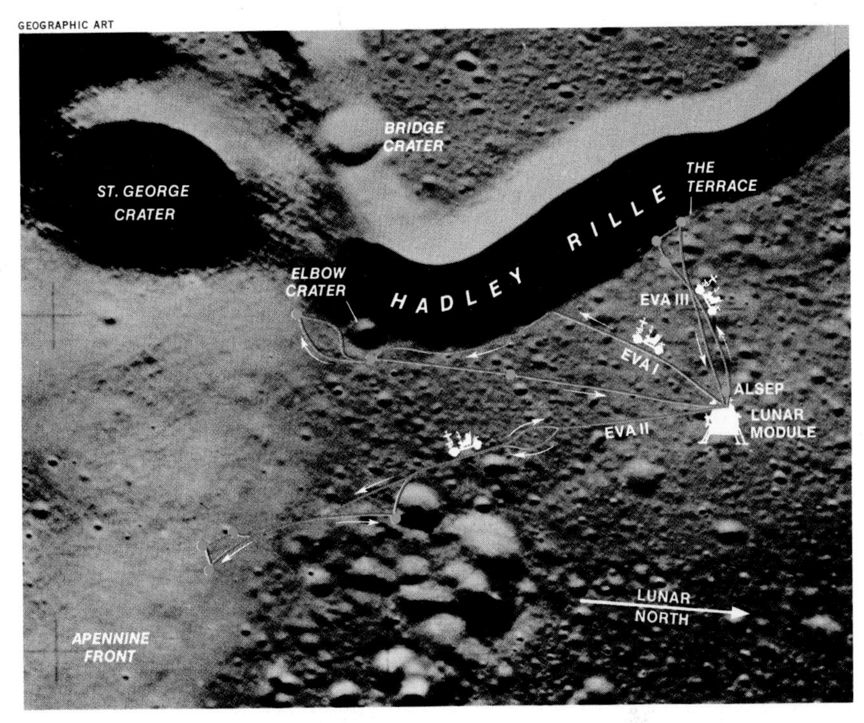

ASTRONAUT *Irwin inspects Rover, the vehicle that carried two Apollo 15 crewmen 17.3 miles in EVA (extravehicular activity) excursions. On the first and third days (EVA I and III on map), the men drove to Hadley Rille. On the second, Rover carried them up Mount Hadley Delta. Intensively trained in geology, the men collected 170.5 pounds of samples. Their lunar module held a larger payload than previous ones, enabling them to take more equipment to the moon and to stay there longer.*

camera, controlled from Houston, showed a great, winding trench with definite stratifications in the opposite wall. The crew worked among the boulders of the near wall, impressing professional geologists with the precision of their descriptions.

When they returned from their first tour to deploy the scientific experiments, Rover's odometer had run up 6.3 miles. At the instrument site Scott had such difficulty drilling into the hard subsurface that blood blisters formed under several of his fingernails. One hole furnished a core sample, and two others held permanently placed sensors to measure subsurface heat flow.

The next day the two visitors made a 7.7-mile round trip well up the slope of the Apennines. The third trip took them to another portion of Hadley Rille, where layering in the wall was particularly evident. From here the great ridges of the Apennines also showed diagonal layering. At the end of the third EVA Scott and Irwin had covered 17.3 miles, had spent a highly productive 18 hours and 33 minutes on the lunar surface, and had recorded and packed 170.5 pounds of rocks.

When they rejoined Al Worden in *Endeavour,* they learned that his busy schedule had been equally productive. With a new, highly precise mapping camera he had photographed at close range vast new areas of the moon, and he had controlled observations by the complex equipment in the SIM (scientific instrument module). On the way home, Worden performed the first space

UNMANNED LUNA 16 *lifts off from the moon's surface, September 20, 1970. First Russian lunar lander to return to earth, the robot spacecraft brought back several ounces of material in a drill-core sample. Computers controlled its 12-day mission; scientists directed the drilling from earth.*

SOVFOTO

DAVID R. SCOTT, NASA

walk outside earth orbit, retrieving film from the panoramic and remote-mapping cameras.

As the *Endeavour* crew sped earthward, they left behind two valuable and active tools: an elaborate surface station hooked up to radio transmitters, and a 78.5-pound satellite in lunar orbit.

Splashdown on August 7 brought a final moment of suspense when one of the three main parachutes collapsed. What looked like quite a hard impact, however, felt less severe inside.

Some of the most unexpected results of the mission came from the heat-flow experiment. Readings of the subsurface temperature showed an average rise of about one degree per foot of depth. Most of the heat flow probably results from radioactivity in the moon's interior. The core from one hole disclosed 58 layers of soil, rendering clues far back into lunar history.

By September 1, 1971, several new conclusions could be drawn. The highland areas of the moon consist of older rocks that contain two to three times as much aluminum as the darker, iron-rich maria. The three-station seismometer network had recorded more than 40 distinct events, including one meteorite impact and two major moonquakes at the extraordinary depth (in comparison with earthquakes) of 450 to 500 miles.

The center of the area where 80 percent of the quakes occurred was beneath a point about 360 miles west of the giant crater Tycho. The surprising depth of the moonquake zone indicated to Dr. Gary V. Latham of Columbia University's Lamont-Doherty Observatory that the moon must be solid, not molten, almost halfway to its center.

Apollo 15 was the first mission to land at the juncture of the moon's dark maria and pale highlands. For Apollo 16, mission planners selected the area around the crater Descartes—7,400 feet higher than the lowest of the landing sites, the Apollo 11 touchdown point.

Astrogeologists were eager to get samples from this possibly volcanic lunar highland, for they believed it could be as much as a billion years older than the maria.

Lift-off for Navy Capt. John Young, Air Force Lt. Col. Charles Duke, and Navy Lt. Comdr. Thomas Mattingly came on April 16, 1972. Among the thousands watching was Russian poet Yevgeny Yevtushenko. He told me he intended to write not about the spectacular lift-off but about his feelings the night before when he beheld the grand aurora of powerful lights pinpointing the tower and rocket and fanning far upward into the sky. "The red tower," he said, ". . . was like an ocean crab that found . . . on the bottom of the sea a rare and unusual white pearl it embraced."

The rarest parts of the pearl, the lunar module *Orion* and the command module *Casper,* reached lunar orbit three days later. Next day, after the two craft had separated and *Casper* went behind the moon for the burn that would put it in a circular orbit above the Descartes region, Ken Mattingly encountered an unexpected problem.

Houston heard the ominous words, ". . . no CIRC," and the men at Mission Control knew they would have to make some agonizing decisions. Mattingly's terse comment referred to a malfunction in the backup guidance system that could

ACTUAL SIZE, NASA (BELOW); DANA SHELTON

CLUES TO EARTH'S ORIGIN *may lie in these moon rocks—prizes of the Apollo 15 landing. Author William R. Shelton (opposite, left) examines a piece of breccia at the Lunar Receiving Laboratory in Houston with Dr. Robin Brett, chief of the Geochemistry Branch. Many scientists believe that the moon formed as gravitation pulled together a cloud of gas and dust. As the moon grew, its gravity increased, and the collision of particles generated enough heat to melt the crust to a depth of 50 to 100 miles. Crystals like those in the anorthosite specimen (left) grew, floated to the surface, and solidified into a crust as the moon gradually cooled. Earth's original crust, now eroded away, possibly formed in a similar manner.*

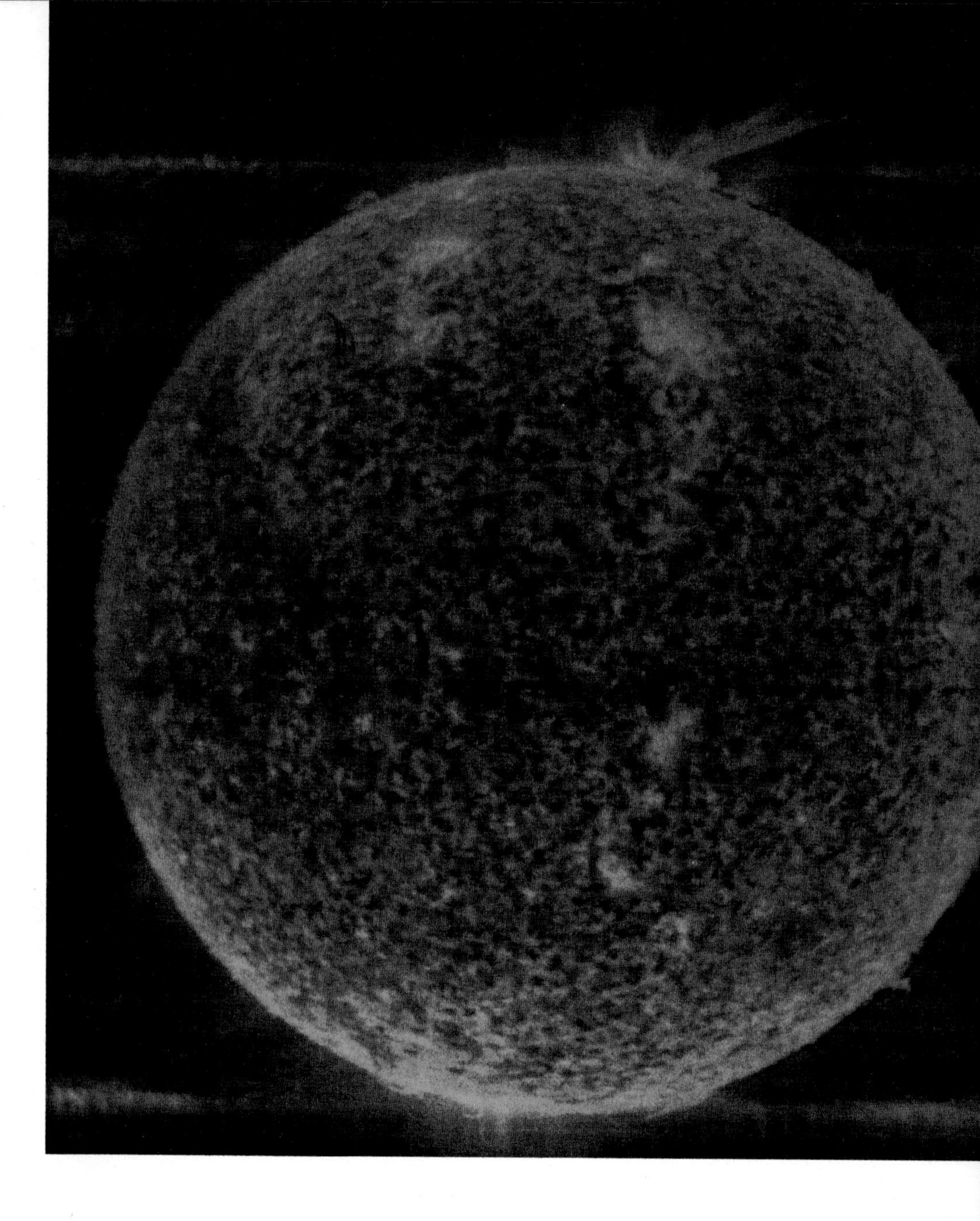

cause oscillation of the engines on which the astronauts depended for their return to earth.

"Looks like we're wiped out already," muttered Deke Slayton, chief of flight crew operations.

Slayton, along with everyone else at Mission Control, knew that even though the primary guidance system checked out, the backup system also had to be operable before the LM could descend. This was because, as with Apollo 13, the LM propulsion system might be needed for the return trip.

Determined experts on the ground bored in on the problem. For four hours the mission hung in precarious suspension. Then, while *Casper* and *Orion* were behind the moon, millions on earth heard the decision before the crew did, when Apollo Control announced that "the oscillations . . . would present no structural hazard to the spacecraft."

Casper's circularization burn behind the moon was nearly perfect. So was *Orion*'s descent burn on the third extra revolution. As Duke read off descent data, John Young found the "flat place" he had said he would be looking for. "Well," he announced, gazing at the rubbled surface, "we don't have to walk far to pick up rocks. . . ."

On their first EVA to Flag crater, the explorers

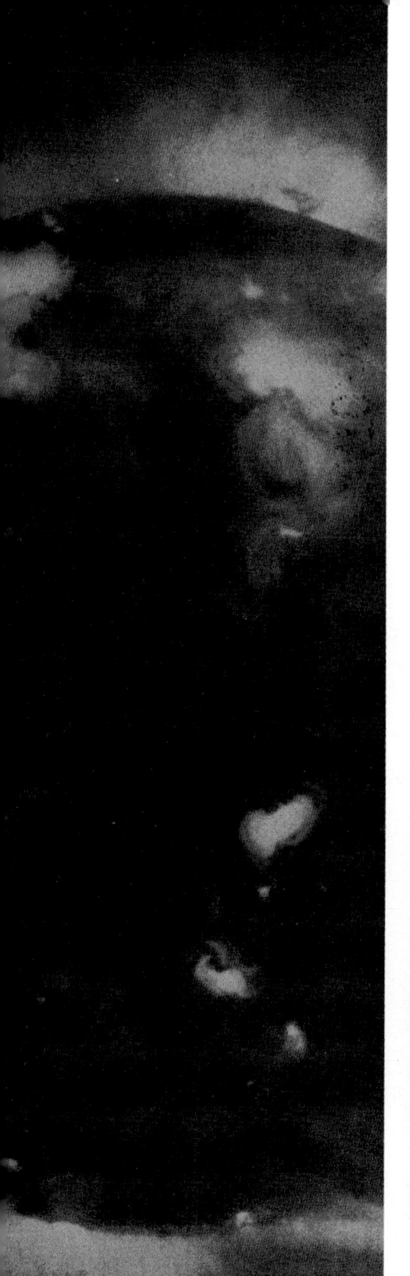

SOLAR FLARE *erupts in a violent burst in a photograph taken June 10, 1973, with the Apollo Telescope Mount during the first manned Skylab mission. The flare, extending 144,000 miles from the sun's surface, fails to show up in the darker image to the right. Exposed at a different wavelength, the second image reveals emissions from solar magnetic regions. At right, Skylab pilot Jack R. Lousma walks in space to load cameras and to deploy a panel that measures micrometeoroid impacts during the second manned flight, July 18 to September 25, 1973.*

NATIONAL AERONAUTICS AND SPACE ADMINISTRATION

found surprisingly few of the crystalline rocks experts had anticipated, but they crossed several distinctly layered ray areas well sprinkled with both rounded and angular rocks.

During the next two days, I watched the longer excursions to Stone Mountain and North Ray crater on three TV sets in a room containing some of the nation's most knowledgeable lunar scientists. Among them was Dr. Harold Masursky of the U. S. Geological Survey. They were overjoyed both with the geologic expertise of Young and Duke and with the greatly improved quality of television which enabled them, at one point, to direct Duke to a promising looking rock. And they were excited about the magnetic readings—up to 313 gammas—indicating a far higher magnetic field than that recorded at any other landing site.

On the third EVA, the explorers somewhat warily approached cavernous North Ray crater, whose sides were so steep they could not see the bottom from a point Young described as being "as close to the edge as I'm going to get."

Later, they were dwarfed by an immense rock far larger than anything ever encountered on the moon. What Dr. Masursky called "a condominium-sized boulder" was estimated by Dr.

Robin Brett to weigh in excess of 18,000 tons.

While gingerly reaching for a sample under a second huge boulder, Duke commented, "Do that in west Texas and you get a rattlesnake. . . ."

Scientists on earth were not in the least disturbed by the apparent rarity of immediately obvious volcanic or "original crust" rocks in the area. "Only the chemistry can tell the whole story," said Dr. Brett. "We're delighted at the wealth of new material we will have to work with."

During excursions totaling nearly 17 miles, the astronauts had accumulated 212 pounds of material, a record for the Apollo series.

With three manned lunar flights canceled by NASA because of budget cuts, the time had now arrived for the final flight. Despite the fact that the target site lay in the bottom of the Taurus-Littrow valley, there was little concern about a possible overshoot in an area that has mountains more than 7,000 feet high. The smooth merger of pilot, computer, and marvelously responsive spacecraft was already solidly proved. What did cause concern was the circumstance that the usual trajectory would require an approach in late 1972 during a solar eclipse. It would place Apollo 17 well within the moon's shadow cone for about nine hours. On-board instruments and systems, flight engineers estimated, could not stand a "cold soak" for that length of time. To avoid this and to permit arrival on target at the desired sun angle, the first night launch was scheduled.

When both initiates and old-time rocket buffs assembled at the Kennedy Space Center, everyone looked forward to a great man-made chandelier that would turn night into day. We were not disappointed: At lift-off, 12:33 a.m., December 7, I turned my eyes away briefly and was startled to see, in the rocket's brilliant glow, many large and small shore birds whirling in the midnight sky.

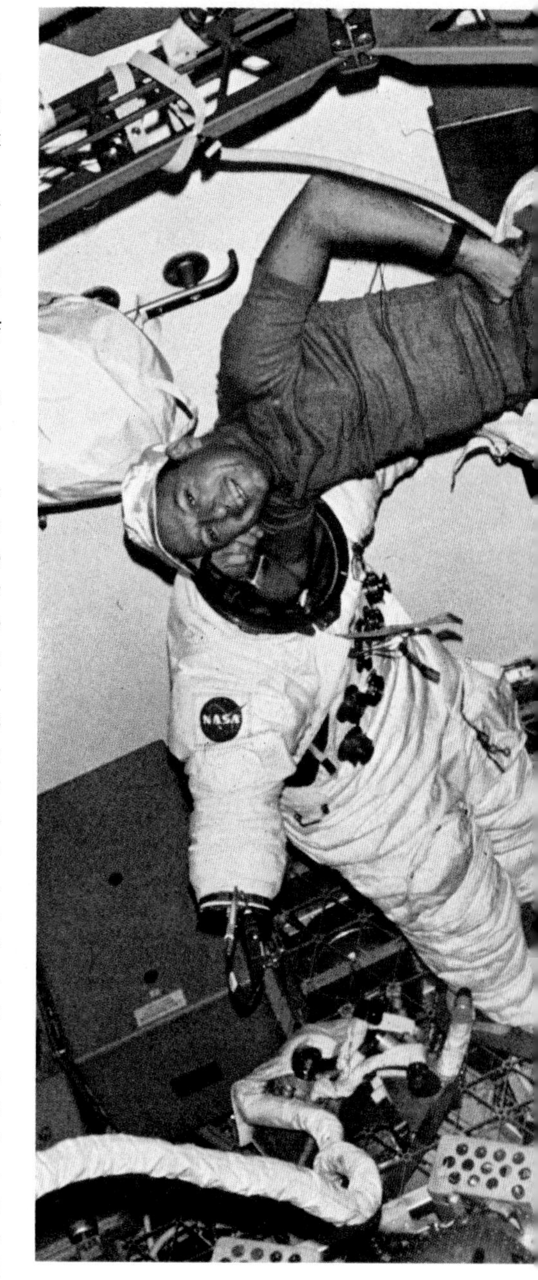

Aboard Apollo 17 were the first geologist in space, Dr. Jack Schmitt, the son of a New Mexico mining geologist; Navy Capt. Gene Cernan, the mission commander, a veteran of both Gemini and Apollo; and command module pilot Ron Evans, who was making his first flight. Also aboard, in contrast to Apollo 11's Spartan inventory of three scientific instruments, were 13 instruments designed to support 26 projects.

After Cernan's perfectly executed touchdown of the lunar module *Challenger* on December 11, Schmitt looked around at what he later called "one of the most majestic panoramas within the experience of mankind. . . . [The site] has the subdued and ancient majesty of a valley whose origins appear as one with the sun."

The first EVA yielded only minor surprises: larger boulders than had been expected from the site photographs and coarse-grained material that initially resembled samples previously taken from the marias. At one point Schmitt took a bad tumble as he maneuvered to scoop up a rock. "Well," he quipped, "I haven't learned to pick up rocks yet, which is very embarrassing for a geologist."

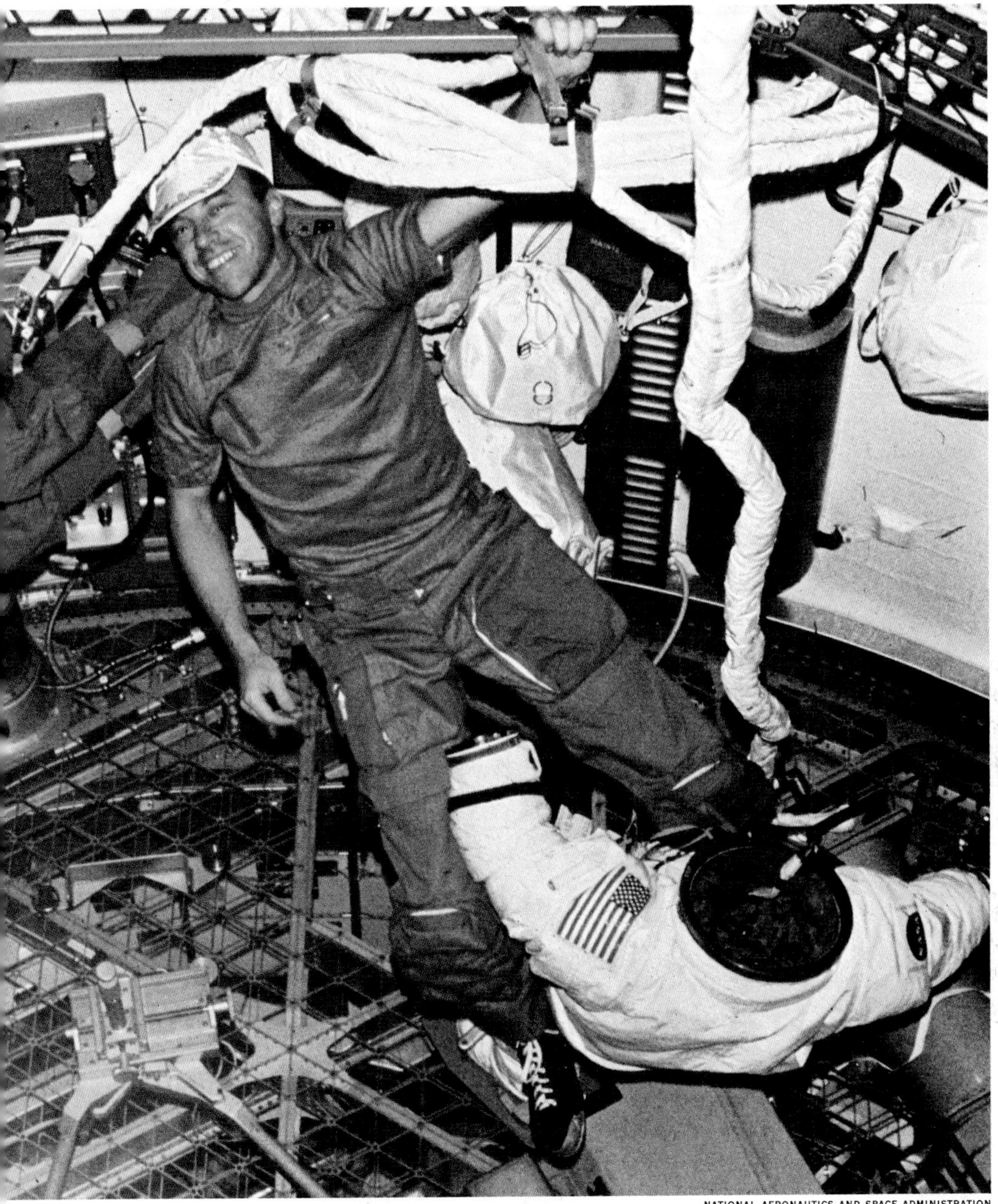
NATIONAL AERONAUTICS AND SPACE ADMINISTRATION

SPACEMEN AT PLAY: *Charles (Pete) Conrad, Jr., and Paul J. Weitz float freely inside the zero-gravity Skylab during the first manned flight. For 28 days—from May 25 to June 22, 1973—they and Joseph P. Kerwin circled the earth every 93 minutes.*

One of the tasks during the second EVA was to repair a fiberglass fender on Rover that had been accidentally broken by Cernan's hammer. The astronauts used a combination of folded lunar maps, tape, and clamps worked out by astronauts and technicians at Mission Control.

A major surprise came during a 12.5-mile round trip to the towering South Massif. Near the crater called Shorty, Schmitt exclaimed happily, "Oh hey! There is orange soil!"

The reason for the enthusiasm on the part of both lunar and earthbound geologists was the possibility that the source of the color was hydrated iron oxide, or rust—possible evidence of the volcanic venting of lunar water from the moon's interior. The fact that the sample was on the surface led Schmitt to think it would be relatively young, perhaps only 10 million years old.

On their third day on the moon, the final EVA to the huge boulders near the Sculptured Hills and the North Massif produced a satisfyingly varied collection of samples.

Before lift-off, Cernan found time to hold up a special rock for the benefit of youngsters of 78 nations invited to Houston for the final moonshot. He called the sample "a very significant rock, composed of many fragments of all sizes and shapes and colors" and told his audience that the Apollo 17 material would be shared with all the countries represented as a symbol to mankind "that we can live in peace and harmony in the future," fitting words from the last man on the moon.

When the *Challenger* crew rejoined with *America* and Ron Evans, they learned that he had spotted two mysterious flashes of light on the lunar surface. These observations, and another flash Schmitt had seen earlier, suggested to some that there might be volcanic venting on the moon.

About halfway back to earth, Evans overcame severe stomach cramps and took a final space walk to retrieve film canisters from two automatic cameras. On December 19, the Apollo 17 crew brought 243 pounds of the moon, plus priceless data, to a perfect splashdown within sight of the carrier U.S.S. *Ticonderoga*.

Analysis of the orange soil proved that it, like the other lunar samples, contained virtually no trace of water. This material, like the green soil retrieved by Apollo 15, was found to be mainly glass. Differing chemically from all the other samples, the orange soil is relatively rich in volatile elements—those that vaporize easily—such as zinc, lead, sodium, and thallium. Although the glass was dated at more than three billion years, its exposure age was estimated at 30 million years.

Early analysis of the lunar materials raised other mysteries. Three theories have been advanced to account for the orange soil: That it was formed by explosive volcanism that ejected droplets of molten lava; that it resulted from a meteorite impact in a pool of lava; or that the intense heat of a meteor crash melted the original lunar rock. More evidence is needed to settle these questions.

Before the final flight, scientists had hoped for rocks both older and younger than any yet found on the moon—perhaps younger than 3.1 billion years. But it soon became apparent that most of the latest samples were clustered in the same relatively narrow time period as the majority of previous samples—roughly between 3.3 and 3.9 billion years. The one notable exception was a four-gram fragment dated at 4.6 billion years.

Based on data from the six widely dispersed Apollo sites, scientists are now in almost general agreement that both the earth and the moon were bombarded by huge chunks of primordial rock between 4.6 billion and 3.9 billion years ago. The last massive bombardment was so great that it obliterated earlier records of violent impacts, and left immense craters and a layer of debris several yards thick on much of the moon's front side.

At the conclusion of the Apollo series, Dr. Rocco A. Petrone, director of the Apollo program, called it "a very romantic era in space exploration," but cautioned that "the book is still being written."

Dr. Petrone was referring to the fact that all over the world, and for many years to come, Apollo's overall treasury of more than 840 pounds of lunar rock and soil samples would undergo intensive scientific analysis. The same attention would be given to the thousands of feet of movie film, video tape, still photographs, and the carefully targeted tests made from lunar orbit by a series of busy command module pilots. Finally, and not least, was the information coming in from the geophysical stations astronauts had left on the moon.

Those who were disappointed that no "grand-slam theorizing" had melded lunar and earth evolution into a single neat package can still suspend judgment. Man's curiosity and sense of wonder might yet be dramatically rewarded in some quiet laboratory. Major surprises are still possible.

From a satellite devoid of life, we are learning how violent was earth's own formation. And we are learning the most fundamental lesson of all: that had not our planet experienced some extra blessing—something beyond diverse minerals and sunlight—it surely would have turned out much like its stillborn sister in space. First our probes of Mars and then of Venus, and now the exploration of the moon, emphatically remind us to revere and protect earth's singular nature and fruition.

ABOVE A SEA OF CLOUDS *Skylab orbits the earth; a sunshade protects the workshop from solar heat. Deployed by Skylab's second crew, the canopy fits over an emergency parasol set up after the spacecraft lost its original shield during launch.*

8 / THE PROMISE OF SPACE: WHERE WILL IT LEAD?

Since the first Sputnik, space exploration has progressed so rapidly that many seemingly extreme predictions have turned out to be conservative. One of the Soviet Union's leading historians of cosmonautics, Evgeny Riabchikov, wrote after his first look at the Salyut space station in 1971, "I gasped and began to laugh. I had to laugh at myself. Decades ago when I had visited Tsiolkovsky and heard his stories of 'ethereal islands,' not only had I failed to believe in these dreams of genius, I had even found them absurd. How I laughed at myself now!"

Several of my own expectations have been dramatically overtaken by events. One day in 1951, when I had just returned from graduate school to teach at Rollins College in central Florida, Dr. Edwin Granberry invited me to join him for a day of surf fishing.

The spot my former teacher selected was a primeval-looking beach that curved gracefully out to a deserted point of land called Cape Canaveral. Behind us were a thick, scrubby undergrowth of semitropical vegetation, a few hammocks of oily green palms, and a variety of snakes and other animal life, including a family of Florida panthers. Man's only mark was an old lighthouse.

As we cast out into the surf for whiting, Ed told me of a rumor that this ecological Eden might become a missile and rocket base.

"Who knows," he said, "maybe within your lifetime you'll see a real live man on the moon."

That's just the way I felt about it, too; possibly sometime late in the 20th century, we might send a man to the moon.

Yet seven years later, at a confidential press briefing only a few miles from that fishing spot, I heard Air Force Maj. Gen. Donald N. Yates make an almost casual announcement: In two days the United States would attempt to rocket an instrument package to the moon from one of the launch platforms that had been erected along the beach.

I was watching when the first reach for the moon failed after 77 seconds of flight. But other

LUNAR COLONY *of the 21st century houses research scientists and visitors from earth, almost a quarter of a million miles away. Spherical craft shuttle passengers to an orbiting space station. In this artist's conception, astronautical engineers survey the moon's pocked surface with the tools of tomorrow.*

STANLEY KUBRICK'S "2001: A SPACE ODYSSEY," METRO-GOLDWYN-MAYER

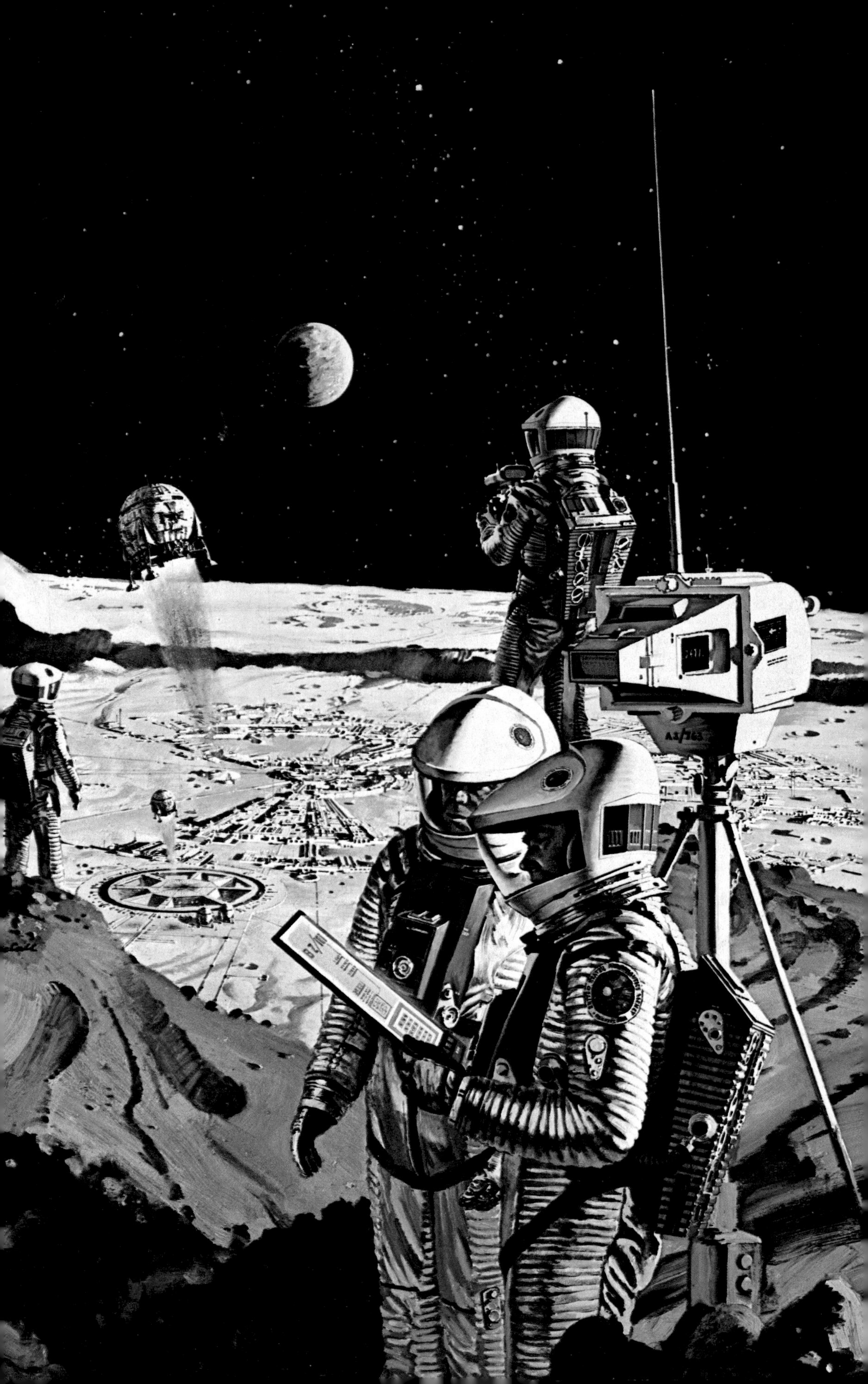

Reusable Shuttle Heralds New Era of Manned Space Missions

Designed for economical earth-orbital missions during the 1980's, the Space Shuttle transportation system (left) climbs skyward after lift-off. A delta-winged passenger and cargo vehicle—the orbiter—rides piggyback on two booster rockets and a larger propellant tank. At an altitude of about 25 miles, the spent boosters will break away and parachute to the ocean; there, a recovery ship will pick them up for later reuse. The expendable propellant tank will fall into a remote area after the Shuttle goes into orbit. With mechanical arms extended, the orbiter (above) retrieves a satellite. Besides repairing or recovering such spacecraft, Shuttle crews will place into earth orbit payloads designed for communications, defense, earth-resources surveys, weather observation, navigation, astronomical research, and pollution control.

rockets succeeded, and successes were repeated, and then in 1969 man arrived on the untrodden lunar surface—far sooner than anyone had predicted even in the late 1950's.

By 1974 man had sustained for 17 years his energetic thrust across a new and hazardous frontier, and had erected a durable bridge into the cosmos. But where does the bridge lead? And why are we going?

John F. Kennedy once used a story by Irish writer Frank O'Connor to illustrate the significance of our foothold in space.

"As a boy," Kennedy related, "he and his friends would make their way across the countryside, and when they came to an orchard wall that seemed too high and too doubtful to try and too difficult to permit their voyage to continue, they took off their hats and tossed them over the wall—and then they had no choice but to follow them."

The space probes sent from earth are, in a sense, the tossed caps that it is now our destiny to follow. Many of the potential rewards are technological; but there are others that may prove of even greater importance. At the time man first placed his footprints on the moon, poet Archibald MacLeish wrote: "To see the earth as it truly is, small and blue and beautiful in that eternal silence where it floats, is to see ourselves as riders on the earth together, brothers on that bright loveliness in the eternal cold—brothers who know now they are truly brothers."

This new view of our planet, appearing so small and fragile in the void, prompted one philosopher to call it "Spaceship Earth," a designation fittingly symbolic of the unique perspective afforded by our trans-atmospheric voyages.

SHUTTLE ORBITER'S NOSE *and under-fuselage (left) glow during re-entry into the earth's atmosphere after a mission in space. The vehicle's heat-shielding exterior, pressurized interior, and lower g-forces will enable passengers to ride in shirt-sleeve comfort. Normally the craft will orbit for about a week, but it could stay aloft for as long as a month if required. Below, the orbiter lands like a jet transport on a conventional runway. The Space Shuttle combines advantages of rockets and airplanes; it will permit greater participation in space travel by astronauts, scientists, engineers, and perhaps even by laymen, for the system could eventually operate as a common carrier.*

NATIONAL AERONAUTICS AND SPACE ADMINISTRATION

ientist, too, we are crossing the
ety of reasons: to extend the ever-
of man, the explorer; to discover
gin, and evolution of the solar
universe; to glean new secrets
s of the atmosphere; to look back
he total environment of earth with
ve instruments ever devised; and
materials, managerial techniques,
es that can be applied to many
human needs. Finally, man is
cosmos out of a powerful urge to
r earth is the only oasis of life in
he universe.

ons of that search intrigue many
s and engineers. A distinguished
observed that the possibility of
ningful histories existing in other parts of the universe has changed the frame of man's religious self-evaluation.

In the Soviet Union, a nation that continues to respond with enthusiasm to the powerful lure of space, a recurring theme interprets the exploration of the cosmos as much more than an extension of the industrial revolution. Before his death in 1935 Konstantin Tsiolkovsky declared that our "man-made environment will be weakened, and may disappear" and suggested that man must be prepared to move, if necessary, to "settlements in space." In the mid-1960's writer Igor Zabelin speculated that "inner motivations are leading mankind to new and unknown shores." In 1971 Evgeny Riabchikov summarized his own view: "The penetration of space is already understood to be not just an isolated experiment but a decision about the destiny and future of humanity. At last

LOWELL GEORGIA

BASE COVER AND PARACHUTE SYSTEM *of a Viking Mars lander rests on a scaffold at an airport in Roswell, New Mexico, while technicians ready it for a balloon-launched deceleration test. Two such robots will land on Mars in 1976 to seek signs of life.*

we have overcome the geocentrism that has dominated the mind of man for thousands of years. . . . The ascent to the cosmos is an irreversible process, the fruit of our entire history."

To exobiologist Carl Sagan of Cornell University, author of *The Cosmic Connection,* the attempt to communicate with other life in the universe is "an idea whose time has come." Dr. Sagan believes that the future will see increased probing for signs of life, although he is concerned about the time lag between earth and the nearest possible star civilization "a few hundred light-years away." Since a light-year is about 6 trillion miles, one-way signals traveling at the speed of light could take as long as 300 years.

"They say 'Hello! How are you?' And 300 years later we say 'Fine!' It's not what you would call snappy dialogue," says Dr. Sagan. But Dr. Sagan believes that messages coming our way—which could be monitored with existing radio telescopes—might be long monologues full of information from other galactic societies.

Both the Soviet Union and the United States have long been interested in such considerations, and most Russian and U. S. scientists favor cooperative ventures in space to avoid duplication and to save money. The most conspicuous example of mutual undertakings is the joint flight of the Soviet Soyuz and the U. S. Apollo, known as the Apollo-Soyuz Test Project (ASTP). This is scheduled for the summer of 1975.

The commander of the Soviet craft will be Col. Alexei A. Leonov, the first man to "walk" in space. His crew mate will be Valeri N. Kubasov. Scheduled to fly in the Apollo craft with Brig. Gen. Thomas P. Stafford are Donald K. (Deke) Slayton and Vance D. Brand.

One purpose of the flight will be to check out a jointly developed docking adapter that will permit rescue operations by crews from either country. Soyuz will go into orbit first, then 7½ hours later Apollo will be launched for rendezvous and docking. The two spacecraft will remain docked for about two days; the crews will exchange visits and conduct joint experiments.

The Soviet Union has already tested Soyuz vehicles much like the ones to be used in ASTP. On September 27, 1973, Lt. Col. Vasili G. Lazarev and Oleg G. Makarov, sent aloft in Soyuz 12, tested the craft's maneuverability during a two-day flight. A second Soyuz, with Maj. Petr I. Klimuk and Valentin Lebedev aboard, was launched on December 18, 1973. Observers have interpreted the eight-day flight as a rehearsal for ASTP, because of the time of launch and orbital altitude.

If the ASTP flight is a success, there may be others. In any event, the main thrust of the United States manned program of the 1970's will be preparation for the new system represented by the Space Shuttle.

The Space Shuttle is expected to be ready in 1979. It will contain enough room for seven people and a cavernous cargo bay 15 feet across and 60 feet long capable of stowing 65,000 pounds of scientific cargo. Its revolutionary feature, however, is its reusability: The delta-winged craft will be able to make a hundred trips or more into orbit, returning each time to earth and landing like an airplane. Pioneer work has already been done by NASA and the U.S. Air Force in flying and landing wingless lifting bodies, sometimes called "flying wedges."

The Shuttle will be able to deliver, repair, reposition, and recover orbiting scientific satellites, and to retrieve film and other data from both manned and unmanned satellites. It will also have a space rescue capability and will be able to dock with other spacecraft. As a vertical ferry it will transport scientists between earth and their space stations, which in the 1980's may be launched in modules and assembled in orbit. Some of these space islands eventually may have rotating elements to create artificial gravity.

If the United States becomes convinced that there are compelling reasons to send astronauts to the moon again, or to one of the asteroids, equipment compatible with the Shuttle probably would be used. It is also possible that crews could revisit Skylab. Such other spacecraft as the unmanned Mars lander could be sent to the moon or to asteroids such as Toro or Icarus. George F. Lawrence of NASA's Langley Research Center in Virginia has said that a modified Viking could place 1,000 pounds of scientific payload at any point on the lunar surface. Instruments landed on the far side would send data to earth via a communications satellite in lunar orbit.

Soviet lunar explorations in the 1970's are expected to extend the Lunokhod robot expeditions and may also be manned. Russia has long studied the possibility of setting up temporary colonies on the moon, perhaps partially underground, where the insulation of the lunar surface material would protect the occupants from temperature extremes and from cosmic rays.

Meteorological and communications satellites will become increasingly serviceable, and their use will continue to increase in the 1970's. Expanded application of laser beams to communications functions might eventually relieve the present overcrowding of radio-frequency bands.

The gradual involvement of other nations in space exploration brought the roster by early 1974 to more than 80. Among them are Great Britain, the Netherlands, Belgium, West Germany, Switzerland, Canada, Australia, Japan, China, and India. Italy has established a launch platform off the coast of Kenya, utilizing the increased velocity

LOWELL GEORGIA (RIGHT); NASA

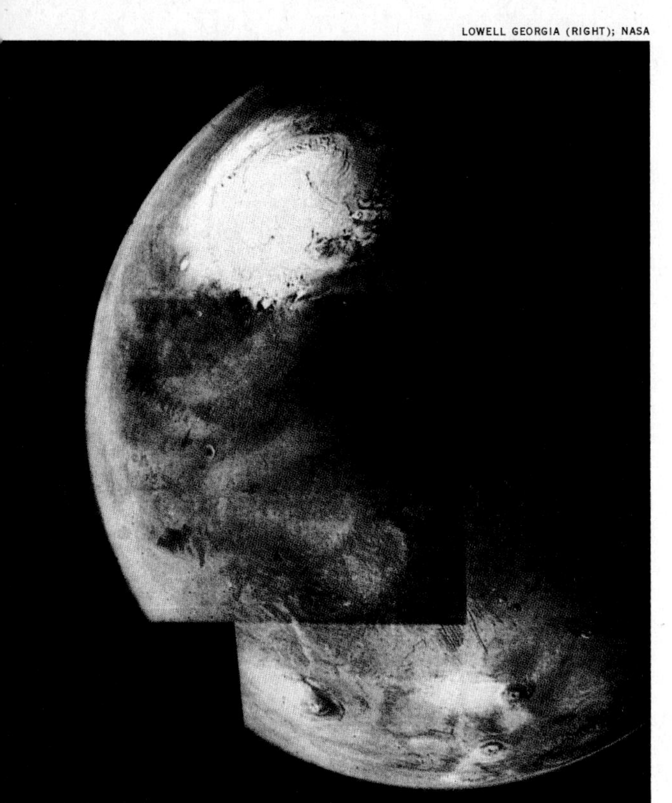

TEN-FOOT *retractable arm on a full-scale model of a Viking lander scoops up a soil sample during a rehearsal at a Colorado test site. In 1975 NASA will launch two such craft for landings on Mars in the summer of 1976. At left, photographs taken by Mariner 9 in 1972 show the red planet's northern ice cap and great volcanoes. To aid the quest for Martian life, Viking carries one of the most complex pieces of machinery ever devised—three automated chemical laboratories, counters for radioactive tracers, tiny ovens, filters, a lamp to simulate Martian sunlight. All these components, compressed into one cubic foot, will analyze measured portions of soil samples. Then dish-shaped antennas will transmit findings to our distant world.*

provided in equatorial zones by the earth's rotation. France has joined with the United States in the Skylab program and with the Soviet Union in the trans-Eurasian communications satellite program. An increasing amount of technical equipment for the Soviet program is manufactured in Czechoslovakia, Poland, East Germany, Rumania, Bulgaria, and Hungary. That even more nations will be involved seems inevitable.

During the 1970's, the United States will launch a variety of satellites—many of them in cooperation with other countries—in the areas of communications, meteorology, navigation, astronomy, and earth-resources study.

By 1978 we should be launching large unmanned earth observatory satellites weighing perhaps as much as 3½ tons. Unfortunately, manned space stations larger than Skylab have been ruled out for the U. S. by the cost of the Space Shuttle program. The Soviet Union, however, is expected to have one or more large space stations in orbit by the end of this decade.

In NASA's tentative plans for the 12 years beginning with 1980, according to Dr. James C. Fletcher, NASA administrator, "We will average three payloads per year to the moon and planets and 57 to earth orbit. That's a ratio of 19 to one in favor of earth orbit during the Shuttle era." This includes manned missions, but it also includes unmanned solar observatories; astronomical observatories, including X-ray telescopes; and advanced satellites in many other areas where smaller, less sophisticated satellites are employed at present.

Says Dr. John F. Clark, Director of Goddard Space Flight Center, "We're slowly putting together a world weather model to help us accomplish a goal of accurate, and I repeat accurate, two-weeks-in-advance weather forecasts."

It seems clear from the 1973 and 1974 launches of several Soviet planetary probes and the U. S. determination to explore planets as far away as Mercury, in one direction, and Jupiter and Saturn, in the other, that both countries will continue to investigate the planets and their moons.

The most ambitious U. S. planetary probes to date are the two Viking landers scheduled for launch in 1975 and arrival at Mars in the summer of 1976. Each spacecraft will contain an orbiter and a lander. When the first Viking gets to Mars, it will brake itself into orbit with a powerful 40-minute burn. Then it will have as many as 30 days before a commitment to land must be made (in case of severe dust storms of the type encountered in 1971 by Mariner 9). However, if everything goes well, the first landing will probably be scheduled for July 4, 1976, on the 200th birthday of the United States.

A sterilized entry system, including a protective aeroshell, a parachute, and braking rockets, will ease each 1,300-pound lander toward the surface.

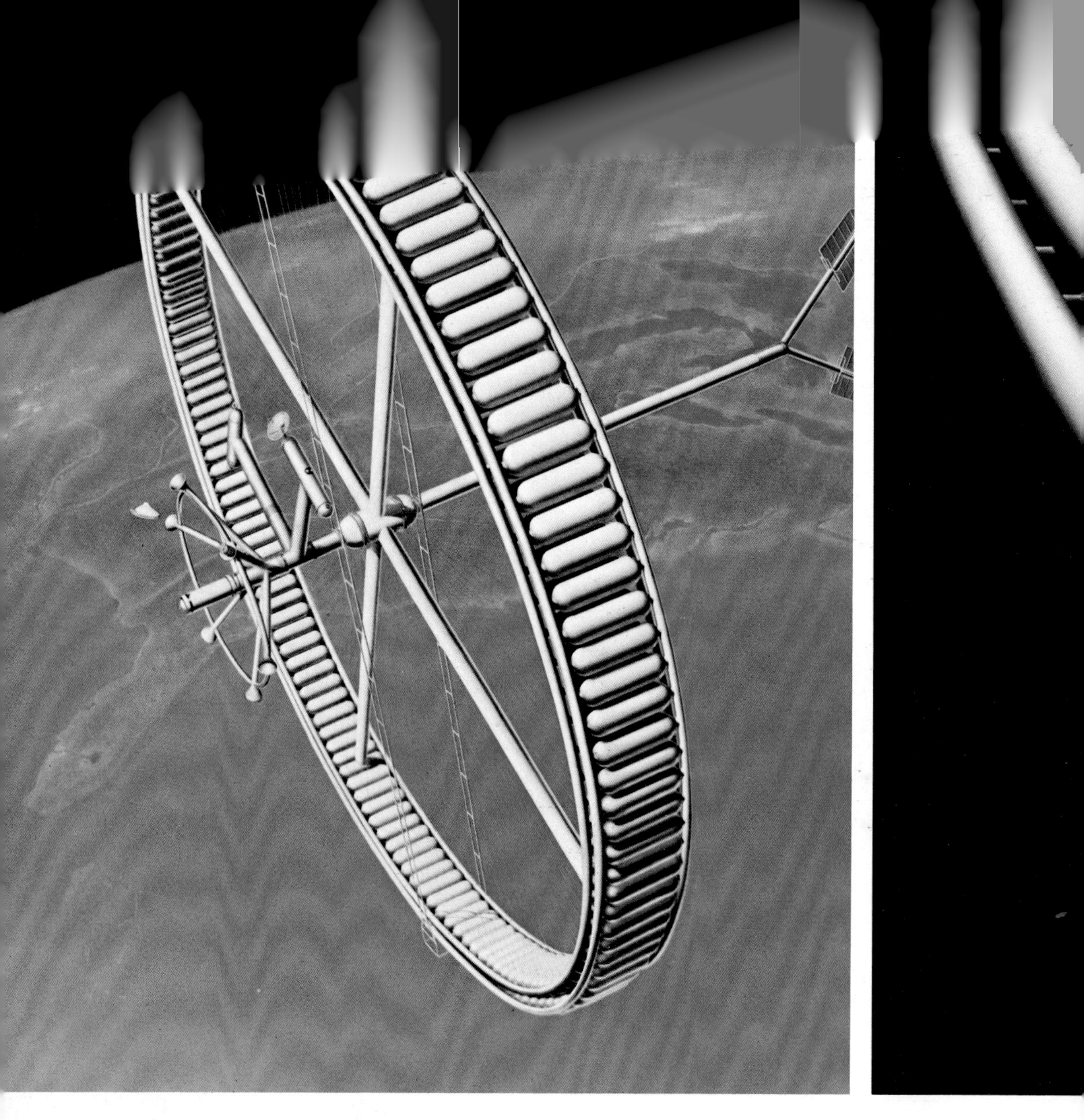

CELESTIAL CITY *of the future wheels silently to create its own artificial gravity 200 miles above Florida. A tiny triangular space liner (left) approaches one of the funnel-shaped docking collars. The cutaway at right reveals a hospital, a computerized library, a dining room, and shuttle tubes. Other sections will hold swimming pools, ice rinks—even a zero-g chamber where the visitor can experience the exhilaration of weightlessness.*

With the help of television cameras in the orbiters, the landers will be able to avoid rugged or hazardous terrain and any slope steeper than 19 degrees. Once a lander is down, a series of complex biological experiments will examine the soil for microscopic life. An ingenious scoop on a ten-foot boom will reach out for soil samples that will be transferred to analytical devices inside the lander. One Viking test will look for evidence of life-forms similar to those on earth. Water containing a mixture of nutrients used by microscopic life on earth will be added to the samples. If the food is consumed, analysis of expelled gases will detect evidence of life processes.

But what if Martian life is different from earth's? Since the atmosphere of Mars allows heavy ultraviolet radiation to reach the surface, life-forms might be protected by silica cells. A second analytic technique will look for micro-organisms from the point of view of Martian conditions. This experiment, instead of adding water to the sample,

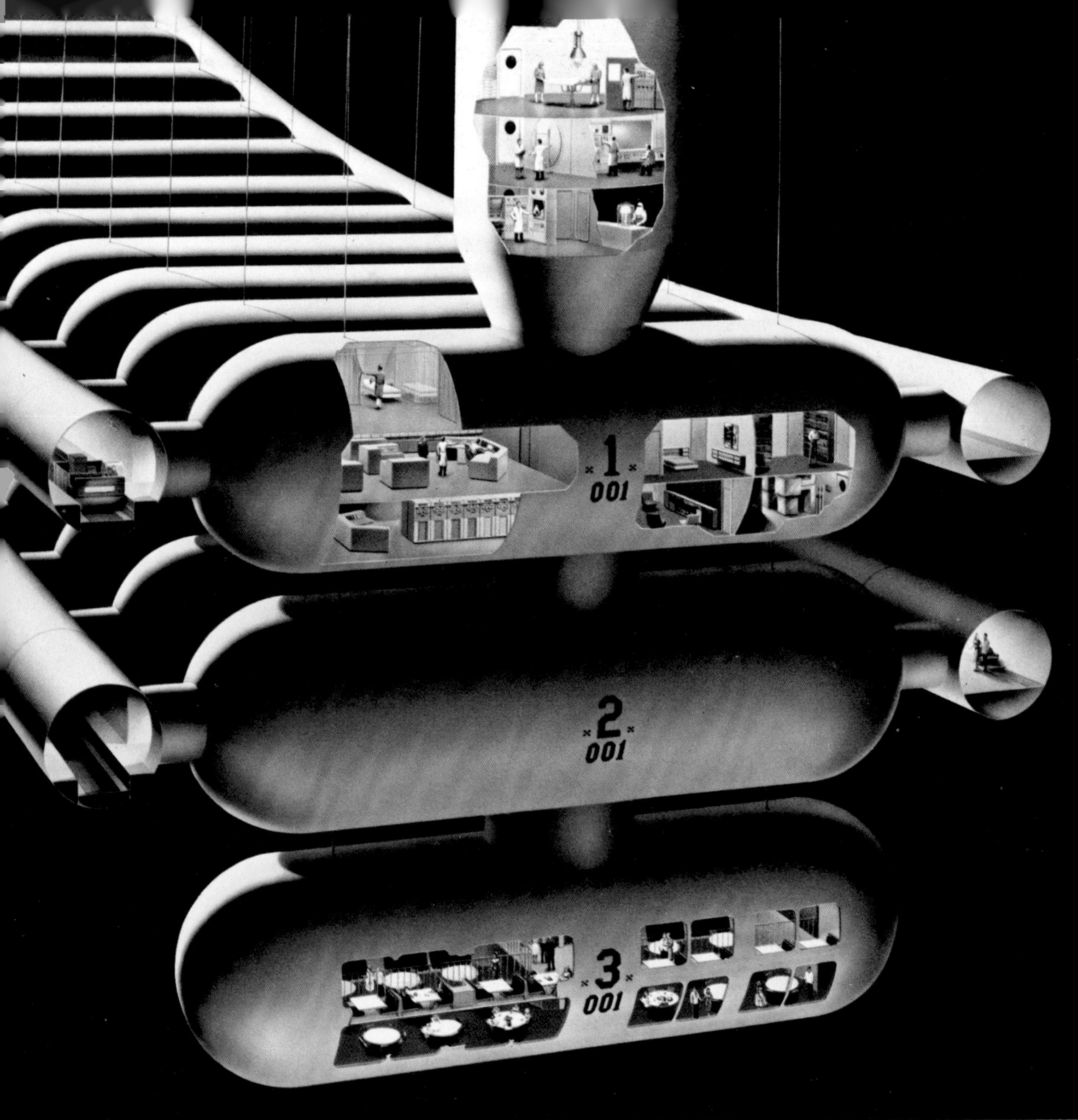

LOCKHEED MISSILES AND SPACE COMPANY

will add radioactive carbon gases and will expose it to simulated Martian sunlight. Still a third experiment, using a small amount of water from earth, will test for a middle range of possible lifeforms: part Martian, part earthlike.

The landers will also take photographs and gather information on soil components, atmospheric conditions, and seismic activity. Some of the data will be sent directly to earth. Other information will be relayed by the orbiters, which are also equipped to analyze the Martian atmosphere and to map the planet's surface.

"For a biologist" says Dr. Gerald A. Soffen, chief scientist of the Viking Project Office, "one of the great scientific opportunities of our time is the search for life on Mars. That question is so important that it could change our whole concept of how life originated."

In late 1974 and in 1975, the U. S. plans to launch a pair of West German solar probes named for the Greek sun god, Helios. The Helios spacecraft will pass so close to the sun that they will be exposed to radiation 11 times greater than that on earth. According to Kurt Heftman, JPL Manager of the Helios project support, the probes will conduct an ambitious series of investigations as they pass by the sun twice each year.

In spite of the cancellation in 1972 for budgetary reasons of the "Grand Tour" of all the outer planets, NASA hopes to use some of the technology to make multiple planetary flights in the late 1970's.

The intrepid Pioneers have already proved they can get through the asteroid belt safely. Now, we will look in faraway places for new answers. The first in-depth study of the outer planets will begin with the 1977 launch of another and more sophisticated spacecraft expected to reach Jupiter in 1979 and Saturn in 1981.

"This will really be an investigation of these planetary systems," says Harris (Bud) Schurmieir of JPL, "because we will also be looking at Jupiter's 12 moons and the 10 moons and rings of Saturn. Near Jupiter, we expect a close flyby of one of Jupiter's four Galilean satellites. In addition, we will sample the interplanetary medium as we go along. If we can keep our instruments going beyond Saturn, we may get data from as far out as 20 AU's [an Astronomical Unit equals about 93 million miles]. If we do, we may get a look at the boundary between the interplanetary medium and the interstellar medium."

Of JPL's far-out projects, including Viking, for which JPL acts in support of the Langley Research Center, Director Bill Pickering says, "The one certainty is that we'll find surprises, maybe very significant surprises. And it's worth pointing out that a negative answer from the first Viking is not necessarily negative in terms of life on Mars."

Two Pioneer spacecraft, one an orbiter and the other carrying four atmospheric probes, may explore Venus in 1978. By 1979, we may send a Viking lander and rover to Mars, and in the future we may investigate one or more comets, including Halley's, which returns in 1985. NASA may send a spacecraft to Uranus in the 1980's, and five or more automated craft could go to the moon in the 1980's to supplement the lunar-orbiting Explorer 49 launched in 1973.

RENDEZVOUS IN LUNAR ORBIT *resupplies a moon base—the glowing cross of lights at lower right. At the far end of a space ferry used to shuttle men and supplies from earth orbit to moon orbit, a stumpy "space tug" grasps cargo containers in its mechanical arms. The tug stacks the canisters upon itself, then descends to the base to unload and return. In this artist's conception, sunlight glints on a solar panel of a space station in orbit above the base. Astronauts tethered to the space station maneuver by individual rocket packs as they survey and photograph the lunar surface.*

FICTION FORECASTS FACT: *Buck Rogers comic strip uses the term "astronaut" in this panel drawn in 1936. The adventure series heralded the Space Age with rocket belts, interplanetary travel, and space stations. At right, in an artist's conception, engineers on a gravity-free platform assemble a craft designed to explore the solar system. Robot arms of a tug guide a fuel tank toward the spaceship. Rockets assembled and launched at such stations would require less propulsion than those launched from earth, where gravity pulls more strongly.*

Deep-probing spacecraft will travel such great distances that panels to utilize solar energy would have to be so large as to be unfeasible. Instead, future Pioneers and Mariners will draw power from nuclear generators. And because of the long times involved—nine years to Pluto, for instance—instruments must be designed to function dependably for long periods.

"You cannot look at the universe as if it is already in your hip pocket," says Daniel Schneiderman, project manager for Mariners 8 and 9, "Unless you're aggressive and determined to be out there, you have no business out there. The human spirit has got to be equal to it."

No specific plans exist for U. S. manned flights to the planets, and some scientists oppose such missions, maintaining that automated vehicles can do the job. But others are convinced that

MARTIN MARIETTA CORPORATION

UNITED
USA
USA
17

man should build nuclear-powered spaceships to visit his neighboring planets in person, and that it is time to begin.

Still other scientists believe the search for life should bypass the solar system and concentrate far out among the stars themselves. But even if we could find a way to travel at the speed of light, it would still take us 4½ years to reach our nearest neighbor star, Alpha Centauri.

So our exploring may be done chiefly from orbital or lunar observatories, with scientists watching and listening across the void for signs of life. Or perhaps we will discover a new way to communicate with or reach other star systems. Most scientists agree that not-too-cold, not-too-hot planets may exist among the galaxies not merely by the dozen but by the hundred, perhaps by the thousand. Is there a "rider" among us who would not like to know whether there are civilizations on other worlds—and how they have fared?

In centuries past, only a chance emissary from some other star system could have provided such an introduction. This is no longer true. In crossing the ramparts of space, man, like the butterfly, has metamorphosed from his former "cocoon" to emerge into a new state of cosmic grace and mobility. He already has rudimentary cosmic wings. And his astonishing technology promises to make true interstellar emissaries of his electronic and mechanical extensions of himself.

In the long march of history, political and economic enthusiasm for exploration and discovery has waxed and waned. The one constant that assured success was man's determination and compulsion. These human attributes continue to exist. They are still challenged by our only non-circumscribed physical frontier. I am more than ever convinced that mankind, having sent its first ambassadors into the cosmos, will follow them in ever-increasing numbers.

EXTRATERRESTRIAL PIONEERS *explore a new-found frontier on Mars. Lifted by gas jets, cameramen make film studies of the dusty terrain. Other crewmen, living in inflated igloos, will chart the surface. A dish antenna beams data to earth, and a landing taxi blasts toward orbit, where a parent ship waits to take the 21st-century explorers home. Of man's future in space, the author says, "not until life moved out to dry land could it see the stars, and life will not be content until it has reached them."*

PAINTING BY ROBERT T. McCALL FROM LIFE MAGAZINE

U.S.A.

ASTRONAUTS AND COSMONAUTS CHALLENGE A VAST NEW FRONTIER

Names of cosmonauts appear in *italics*.

Edwin E. Aldrin, Jr.

William A. Anders

Neil A. Armstrong

Alan L. Bean

Eugene A. Cernan

Michael Collins

Charles Conrad, Jr.

L. Gordon Cooper, Jr.

R. Walter Cunningham

Georgi T. Dobrovolsky

Owen K. Garriott

Edward G. Gibson

John H. Glenn, Jr.

Victor V. Gorbatko

Richard F. Gordon, Jr.

Virgil I. Grissom

Valeri N. Kubasov

Vasili G. Lazarev

Valentin Lebedev

Alexei A. Leonov

Jack R. Lousma

James A. Lovell, Jr.

William R. Pogue

Pavel R. Popovich

Stuart A. Roosa

Nikolai N. Rukavishnikov

Walter M. Schirra, Jr.

Harrison H. Schmitt

Thomas P. Stafford

John L. Swigert, Jr.

Valentina V. Tereshkova

Gherman S. Titov

Vladislav N. Volkov

Boris V. Volynov

HUMANS VENTURED *into space 51 times from April 1961 through December 1973. This chronological list includes launch dates and members of each mission:* VOSTOK 1, *April 12, 1961: Gagarin, first in space, one orbit.* MERCURY 3, *May 5, 1961: Shepard, first American in space, suborbital flight.* MERCURY 4, *July 21, 1961: Grissom completes suborbital flight, capsule accidentally sinks.* VOSTOK 2, *August 6, 1961: Titov, 17 orbits.* MERCURY 6, *February 20, 1962: Glenn, 3 orbits.* MERCURY 7, *May 24, 1962: Carpenter, 3 orbits.* VOSTOK 3, *August 11, 1962, Nikolayev;* VOSTOK 4, *August 12, 1962, Popovich: two craft pass within 3 miles of each other.* MERCURY 8, *October 3, 1962: Schirra, 6 orbits.* MERCURY 9, *May 15, 1963: Cooper, 22 orbits.* VOSTOK 5, *June 14, 1963, Bykovsky;* VOSTOK 6, *June 16, 1963, Tereshkova, first woman in space: two craft pass 3 miles apart.* VOSKHOD 1, *October 12, 1964: Komarov, Feoktistov, and Yegorov, first three-man craft.* VOSKHOD 2, *March 18, 1965: Leonov, first space walk, Belyayev pilots.* GEMINI 3, *March 23, 1965: Grissom and Young, first U. S. two-man craft.* GEMINI 4, *June 3, 1965: McDivitt pilots, White walks in space.* GEMINI 5, *August 21, 1965: Cooper and Conrad orbit 8 days in sustained-flight test.* GEMINI 7, *December 4, 1965: Borman and Lovell stay aloft 2 weeks.* GEMINI 6, *December 15, 1965: Schirra and Stafford approach within a foot of Gemini 7.* GEMINI 8, *March 16, 1966: Armstrong and Scott dock with target vehicle.* GEMINI 9, *June 3, 1966: Stafford and Cernan make varying rendezvous approaches and conduct intricate maneuvers with target vehicle.* GEMINI 10, *July 18, 1966: Young and Collins dock and maneuver with target vehicle.* GEMINI 11, *September 12, 1966: Conrad and Gordon reach 853-mile apogee.* GEMINI 12, *November 11, 1966: Lovell pilots, Aldrin works outside capsule for 5½ hours.* SOYUZ 1, *April 22, 1967: Komarov dies after re-entry when spacecraft parachutes fail.* APOLLO 7, *October 11, 1968: Schirra, Eisele, and Cunningham fly first U. S. three-man craft.* SOYUZ 3, *October 26, 1968: Beregovoi draws near unmanned Soyuz 2.* APOLLO 8, *December 21, 1968: Borman, Lovell, and Anders, first to orbit the moon.* SOYUZ 4, *January 14, 1969, Shatalov;* SOYUZ 5, *January 15, 1969, Volynov, Yeliseyev, and Khrunov: two craft dock and Yeliseyev and Khrunov transfer to Soyuz 4 for return to earth.* APOLLO 9, *March 3, 1969: McDivitt, Scott, and Schweickart test lunar module and*

Pavel I. Belyayev · Georgi T. Beregovoi · Frank Borman · Valeri F. Bykovsky · M. Scott Carpenter · Gerald P. Carr
Charles M. Duke, Jr. · Donn F. Eisele · Ronald E. Evans · Konstantin P. Feoktistov · Anatoli V. Filipchenko · Yuri A. Gagarin
Fred W. Haise, Jr. · James B. Irwin · Joseph P. Kerwin · Yevgeni V. Khrunov · Petr I. Klimuk · Vladimir M. Komarov
Oleg G. Makarov · Thomas K. Mattingly II · James A. McDivitt · Edgar D. Mitchell · Andriyan G. Nikolayev · Victor I. Patsayev
Russell L. Schweickart · David R. Scott · Vitali I. Sevastyanov · Vladimir A. Shatalov · Alan B. Shepard, Jr. · Georgi S. Shonin
Paul J. Weitz · Edward H. White II · Alfred M. Worden · Boris B. Yegorov · Alexei S. Yeliseyev · John W. Young

take 1,353 photographs of earth. APOLLO 10, *May 18, 1969: Stafford and Cernan maneuver lunar module to within 9 miles of moon's surface, Young pilots command module.* APOLLO 11, *July 16, 1969: Armstrong and Aldrin make first moon landing, Collins remains in orbit.* SOYUZ 6, *October 11, 1969, Shonin and Kubasov;* SOYUZ 7, *October 12, 1969, Filipchenko, Volkov, and Gorbatko;* SOYUZ 8, *October 13, 1969, Shatalov and Yeliseyev: three craft fly in formation, men test-weld in space.* APOLLO 12, *November 14, 1969: Conrad and Bean explore lunar surface while Gordon orbits.* APOLLO 13, *April 11, 1970: Lovell, Swigert, and Haise return safely from aborted moon-landing mission after oxygen tank bursts.* SOYUZ 9, *June 1, 1970: Nikolayev, Sevastyanov, 18 days in orbit.* APOLLO 14, *January 31, 1971: Shepard and Mitchell explore lunar highlands, Roosa orbits.* SOYUZ 10, *April 22, 1971: Shatalov, Yeliseyev, and Rukavishnikov dock with Salyut space station.* SOYUZ 11, *June 6, 1971: Dobrovolsky, Volkov, and Patsayev conduct successful work mission inside Salyut, die on re-entry when pressure leak causes cabin decompression.* APOLLO 15, *July 26, 1971: Scott and Irwin explore Hadley Rille, slope of Apennine Mountains in the lunar Rover. Worden conducts experiments in orbit.* APOLLO 16, *April 16, 1972: Young and Duke explore the Descartes highlands in an improved vehicle, as Mattingly works in orbit.* APOLLO 17, *December 7, 1972: Cernan and Schmitt explore the Taurus-Littrow valley in a lunar Rover; Evans performs orbital geophysical experiments.* SKYLAB 2 *(first manned), May 25, 1973: Conrad, Kerwin, and Weitz repair the damaged orbital workshop and conduct medical, solar, and earth-resources experiments for 28 days.* SKYLAB 3 *(second manned), July 28, 1973: Bean, Garriott, and Lousma perform detailed medical, solar, and earth-resources experiments during their 59 days in orbit.* SOYUZ 12, *September 27, 1973: Lazarev and Makarov test modifications to the spacecraft and space suits.* SKYLAB 4 *(third manned), November 16, 1973: Carr, Gibson, and Pogue study the sun and comet Kohoutek, and conduct medical and earth-resources experiments during their record-breaking 84 days in orbit. Using the Apollo Telescope Mount, Gibson makes a unique recording of a solar flare, including its initial flash phase.* SOYUZ 13, *December 18, 1973: Klimuk and Lebedev further test the Soyuz equipment in preparation for a joint U. S.-U.S.S.R. mission in 1975.*

ASTRONAUTS: NASA; COSMONAUTS: SOVFOTO, NOVOSTI

INDEX

Boldface indicates illustrations; *italic* refers to picture captions

ADDITIONAL READING

For additional reading, you may wish to refer to these NATIONAL GEOGRAPHIC articles and to check the Cumulative Index for related material.

"How We Mapped the Moon," by David W. Cook, Feb. 1969. "Footprints on the Moon," by Hugh L. Dryden, March 1964. "Telephone a Star," by Rowe Findley, May 1962. "Cape Canaveral's 6,000-mile Shooting Gallery," Oct. 1959; "Exploring Tomorrow With the Space Agency," July 1960; "Reaching for the Moon," Feb. 1959, all by Allan C. Fisher, Jr. "The Sun," by Herbert Friedman, Nov. 1965. "The Making of an Astronaut," by Robert R. Gilruth, Jan. 1965. "Space Satellites, Tools of Earth Research," by Heinz Haber, April 1956. "Apollo 14: The Climb Up Cone Crater," by Alice J. Hall, July 1971. "The Flight of *Freedom 7,*" by Carmault B. Jackson, Jr., M.D., Sept. 1961. "How Man-Made Satellites Can Affect Our Lives," by Joseph Kaplan, Dec. 1957. "The Earth From Orbit," by Paul D. Lowman, Jr., Nov. 1966. "The Laser's Bright Magic," by Thomas Meloy, Dec. 1966. "Exploring our Neighbor World, the Moon," by Donald H. Menzel, Feb. 1958.

"First Color Portraits of the Heavens," by William C. Miller, May 1959. "Surveyor: Candid Camera on the Moon," by Homer E. Newell, Oct. 1966. "A Most Fantastic Voyage," by Lt. Gen. Sam C. Phillips, USAF, May 1969. "Mars, a New World to Explore," by Carl Sagan, Dec. 1967. "Mariner Scans a Lifeless Venus," May 1963; "Robots to the Moon," Oct. 1962, both by Frank Sartwell. "The Pilot's Story: Astronaut Shepard's Firsthand Account of His Flight," by Alan B. Shepard, Jr., Sept. 1961. "The Moon Close Up," by Eugene M. Shoemaker, Nov. 1964. "Our Earth as a Satellite Sees It," by W. G. Stroud, Aug. 1960. "John Glenn's Three Orbits in *Friendship 7,*" by Robert B. Voas, June 1962.

"Apollo 15: To the Mountains of the Moon," Feb. 1972; "Countdown for Space," May 1961; "Giant Comet Grazes the Sun," Feb. 1966; "Historic Color Portrait of Earth from Space," Nov. 1967; "The Moon, Man's First Goal in Space," Feb. 1969; "Of Planes and Men," Sept. 1965; "Space Rendezvous, Milestone on the Way to the Moon," April 1966; "Tracking America's Man in Orbit," Feb. 1962, all by Kenneth F. Weaver. "America's 6,000-mile Walk in Space," Sept. 1965. "Extraordinary Photograph Shows Earth Pole to Pole," Feb. 1965. "First Explorers on the Moon—The Incredible Story of Apollo 11," December 1969; "School for Space Monkeys," May 1961.

NATIONAL GEOGRAPHIC articles published since the 1972 edition of the book: "Journey to Mars" and "The Search for Life on Mars," by Kenneth F. Weaver, Feb. 1973. "Apollo 16 Brings Us Visions From Space," Dec. 1972. "Summing Up Mankind's Greatest Adventure," Sept. 1973.

Composition for this edition of *Man's Conquest of Space* by National Geographic's Phototypographic Division, Carl M. Shrader, Chief; Lawrence F. Ludwig, Assistant Chief. Printed and bound by Fawcett Printing Corp., Rockville, Md. Color separations by Colorgraphics, Inc., Beltsville, Md.

NATIONAL GEOGRAPHIC SOCIETY

WASHINGTON, D. C.

Organized "for the increase and diffusion of geographic knowledge"

GILBERT HOVEY GROSVENOR
Editor, 1899-1954; President, 1920-1954
Chairman of the Board, 1954-1966

THE NATIONAL GEOGRAPHIC SOCIETY is chartered in Washington, D. C., in accordance with the laws of the United States, as a nonprofit scientific and educational organization for increasing and diffusing geographic knowledge and promoting research and exploration. Since 1890 the Society has supported 1,005 explorations and research projects, adding immeasurably to man's knowledge of earth, sea, and sky. It diffuses this knowledge through its monthly journal, NATIONAL GEOGRAPHIC; more than 50 million maps distributed each year; its books, globes, atlases, and filmstrips; 30 School Bulletins a year in color; information services to press, radio, and television; technical reports; exhibits from around the world in Explorers Hall; and a nationwide series of programs on television.

NATIONAL GEOGRAPHIC MAGAZINE

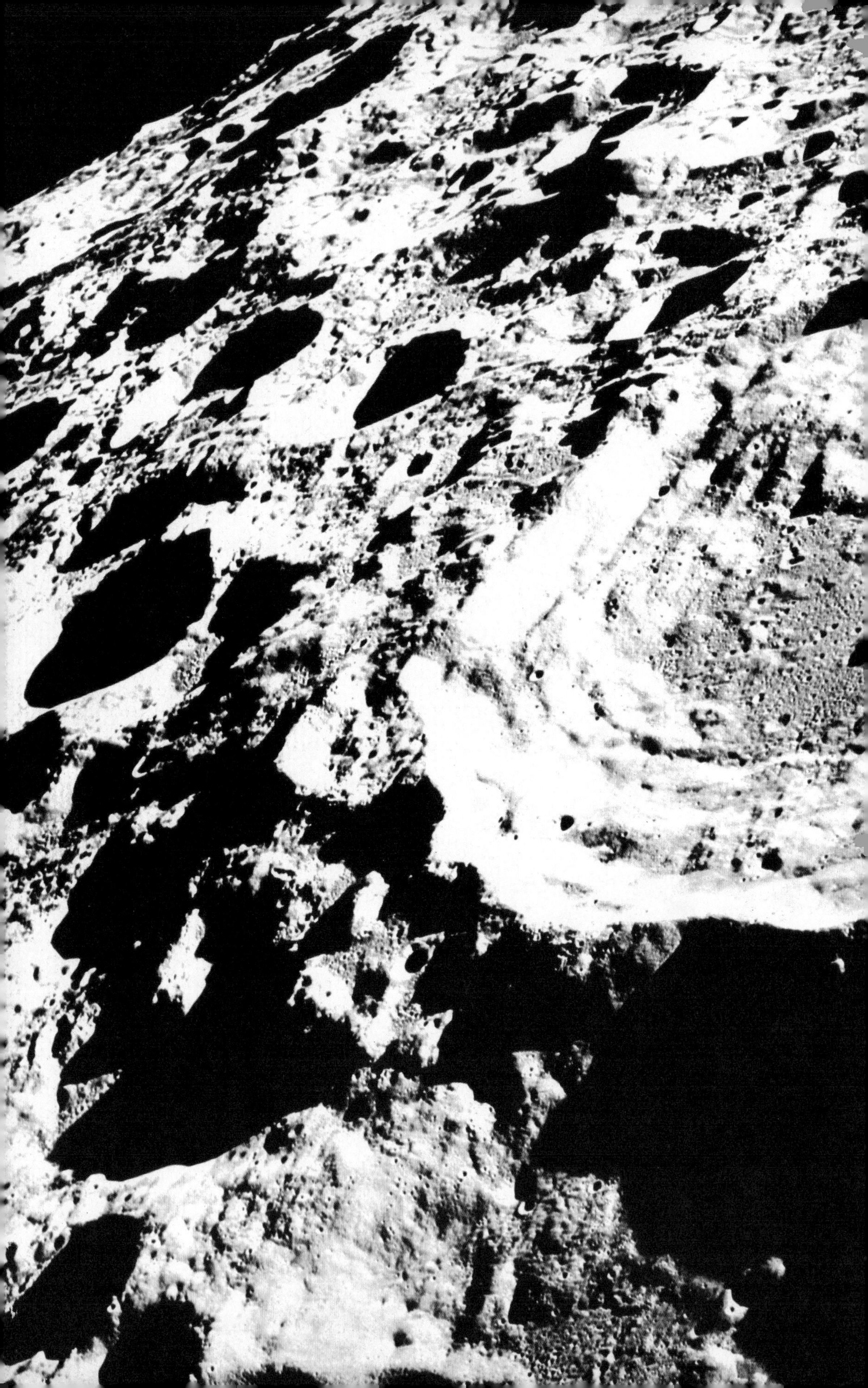